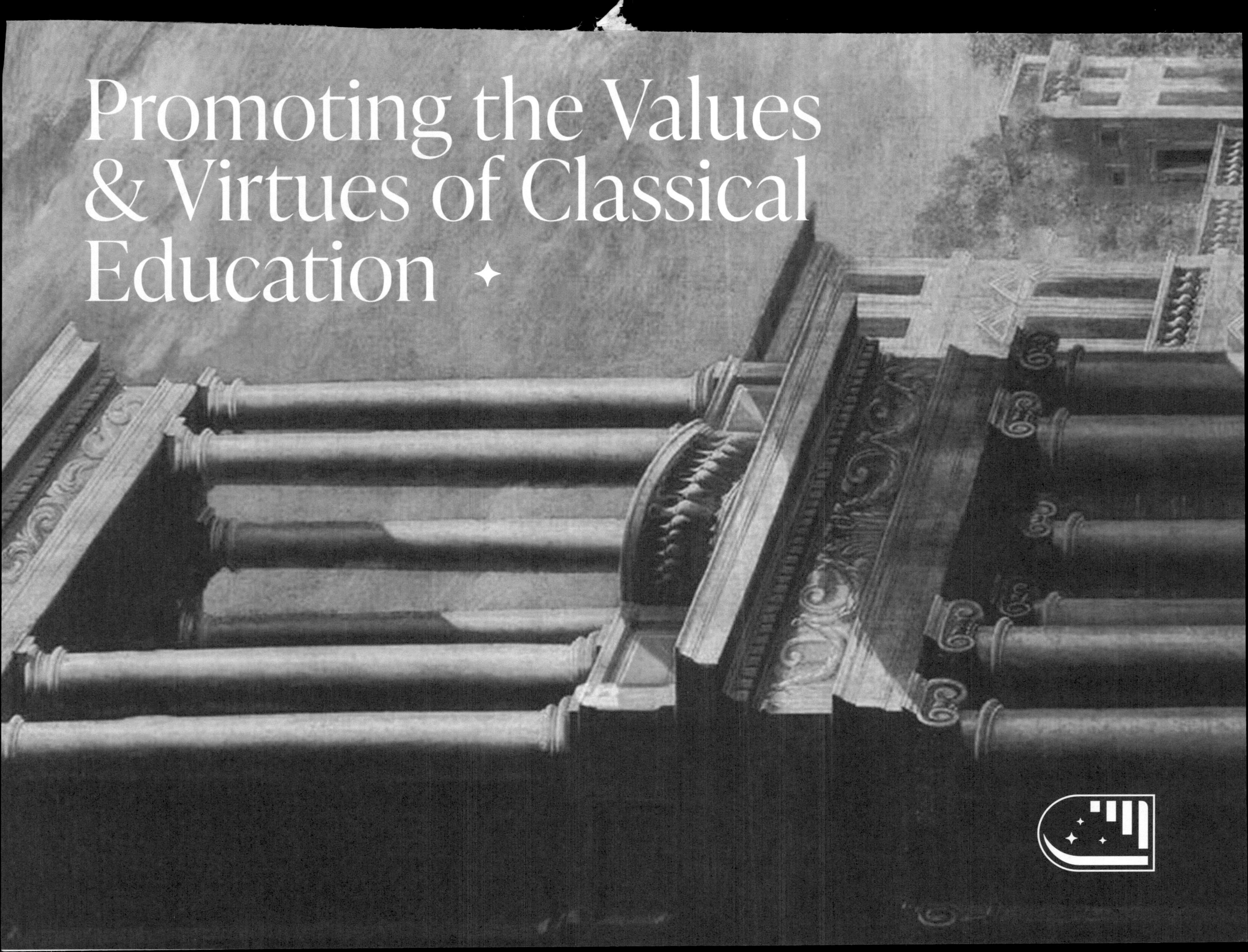
Promoting the Values
& Virtues of Classical
Education ✦

Fundamentals of Engineering

Thales Press Raleigh, North Carolina

This book is the property of:

....

....

....

....

Luddy Industrial Arts: Fundamentals of Engineering

For tests, answer keys, and supplementary materials, please email us at thalespress@thalesacademy.org.

The cover photo is a 3-D printer, created by Austin Williams as part of his senior project at Thales Academy Rolesville Junior High / High School.

NOTEBOOK NO. ____________________

CONTINUED FROM NOTEBOOK NO. ________ CONTINUED TO NOTEBOOK NO. ________

ASSIGNED TO:

NAME: ____________________

SIGNATURE: ____________________ DATE: ________

DATE ISSUED: ____________________ BY: ____________

SIGNATURE: ____________________

PHONE: ____________________

EMAIL: ____________________

SCHOOL: ____________________

CLASS: ____________________

ADDRESS: ____________________

____

CITY: __________ STATE: __________ ZIP: __________

DATE NOTE BOOK COMPLETED: __________ NUMBER OF PAGES FILLED IN: __________

NOTES:

____

____

____

Table of Contents

Photo from NASA

Module I
Engineering Fundamentals

TOPICS

01 Safety & Responsibility

02 Engineering Design Cycle

03 Engineering Lettering

04 Isometric Drawing

05 Orthographic Drawing

06 Metrology

07 Packaging

08 Properties of Materials

09 Forces

CAPTIVEAIRE SYSTEMS / RALEIGH, NORTH CAROLINA

LESSON

Safety & Responsibility

ROADMAP

- Understand all rules regarding classroom safety, personal responsibility, and classroom etiquette.
- Familiarize yourself with your classroom and where all the safety equipment is stored.

Safety & Responsibility

What is it and why is it important?

Safety is the most important thing engineers must consider in their daily jobs. Any tool, when used improperly, can become a danger to you and those around you. That is why we are going to explore the common tools you will use throughout this curriculum and the proper way to use them. It is important to remain safe when working on any project, so reference this lesson often and always ask your teacher if you have any questions or concerns.

Classroom Safety

1. Do not run in the classroom.
2. Do not horse around.
3. Remain seated unless you are required to get up.
4. No food or drink in the classroom.
5. Do not touch any equipment without permission.
6. Treat other students' project work with respect. Look but don't touch.
7. When you enter the classroom, put on a pair of safety glasses, then sit down as soon as possible. The teacher will instruct students to wear gloves or dust masks when required.
8. Keep your noise level to a minimum. The teacher should be able to be heard at all times to prevent accidents.
9. Push in stools whenever possible so people can move around freely. Stools can be a dangerous tripping hazard.
10. No bookbags in the classroom. Students should only require a pencil, calculator, iPad, and their workbook.
11. Students are expected to always wear closed toe shoes and pants in the classroom.

Personal Responsibility

1. Remind others to follow the safety procedures if you see them doing anything unsafe.
2. If you spill something, clean it up immediately.
3. At the end of class, everyone must actively help with cleanup.
4. Put away any tools you used, use the broom and vacuum cleaner to clean up any sawdust, and wipe down all tables.
5. Take care of the tools as if they were your own. Absolutely no playing with tools.
6. Always push in your stool before you leave the class.

Classroom Etiquette

1. Give all of your attention to the teacher when the teacher is speaking. This means:

    a. No talking while the teacher is talking.

    b. No using tools or working on projects while the teacher is talking.

    c. Raise your hand if you have a question or comment.

Where to find it? / Safety & Responsibility

Use the space provided to list the important items in your classroom and note where they are.

Where are the safety glasses? Where is your kit? Where is the fire extinguisher?

Draw the Classroom / Safety & Responsibility

Use the space provided to carefully draw the floor plan of your classroom. Make sure to label everything that you listed on the previous page.

THE PARTHENON / ATHENS, GREECE

Photo by Josh Stewart

LESSON

Engineering Design Cycle

ROADMAP

- Learn the Engineering Design Cycle.
- Learn how to apply the Engineering Design Cycle to everyday problems.

Engineering Design Cycle

What is it and why is it important?

The engineering design cycle provides a framework for engineers to invent creative solutions to problems and improve current technologies. This important engineering concept has been in use for thousands of years, but became the formal cycle that you will learn about in today's lesson in the mid-20th century.

FIGURE 2.1
ENGINEERING DESIGN CYCLE

How does the engineering design cycle work?

The engineering design cycle has six steps.

What should you do if you have a lamp that doesn't generate light?

1: Ask Questions

Ask yourself the question, "Why doesn't my lamp generate light?" This question is more commonly known as your **problem statement.**

Problem Statement: A short, clear explanation of an issue which you want to solve.

Ask questions related to the **criteria** of the problem you are trying to solve. For example, ask questions about why the lamp doesn't generate light.

Criteria: A list of the requirements a solution must meet to be considered successful.

Ask questions about the **Constraints** of solving this problem. For example, one constraint is that you can't take the lamp apart to repair it.

Constraints: A list of restrictions that limit or control what you can do.

Vocabulary

Engineering Design Cycle
A process used to break down and find the best solution to the problem at hand.

Ask Questions
Ask questions about why something doesn't work properly or about the problem you are trying to solve.

Problem Statement
A short, clear explanation of a problem you want to solve.

Criteria
A list of the requirements a solution must meet to be considered successful.

Constraints
A list of things that limit or control what you can do.

Imagine
Imagine possible solutions to the problem.

Make a Plan
Using your possible solutions, put a plan together in the form of a checklist to follow.

Create
These are steps to test your possible solutions. This is the first prototype of a possible solution.

Test
The steps needed to test the prototype to find the problem.

Improve
If you think of any other possible solutions when testing, you can improve the checklist and steps to solve the problem.

2: Imagine

Imagine possible solutions to the problem.

What are some possible reasons my lamp doesn't generate light?

For example:

Is my lamp plugged in?

Is the lightbulb broken?

Does the outlet have power?

Is the lamp broken internally?

Does my house have power?

3: Make a Plan

Using your possible solutions, put a plan together in the form of a checklist to follow.

Tip: start with the easiest solutions first.

Checklist for why my lamp doesn't work:

- ☐ Does my house have power?
- ☐ Is the lamp plugged in?
- ☐ Does the outlet have power?
- ☐ Is the light bulb broken?
- ☐ Is the lamp broken inside?

4: Create

Using the checklist, create a numbered list of steps that you can follow to test the reasons your lamp may not generate light. This checklist should lead you to possible solutions.

Examples include the following:

1. "Does my house have power?" Check if the ceiling lights work. If they don't then the house has no power. If the lights do work, the house has power.
2. "Is the lamp plugged in?" Look to make sure the cord is plugged in. If not, then plug it in and try the lamp. If it still does not work, then continue with the next step.
3. "Does the outlet have power?" Check to see if the outlet has power by plugging in something you know works. If it works, then move on. If it doesn't work then the outlet has no power, so move the lamp to a working outlet. If the lamp still does not work, then continue with the next step.
4. "Is the bulb broken?" Replace the bulb in the lamp with a new one. If the lamp works, then the problem is solved. If the lamp still doesn't work, then continue with the next step.
5. "Is the lamp broken internally?" If the lamp has not worked after all the tests, then the lamp is broken internally, and by the constraints outlined at the beginning of the process, the lamp can not be taken apart to be repaired. At this point, you must decide whether to have the lamp repaired by an electrician or if you should dispose of the lamp.

Among his many accomplishments, Thomas Edison invented the first electric lightbulb. The process, though, was not without its challenges, as Edison reportedly went through thousands of different materials and designs until he tried out bamboo as the filament (i.e., the material that burns and thereby provides light).

This bamboo-filament burned for a long period of time, used a small amount of electricity, and was proven to be safe for use in homes and offices.

Edison remarked of the process, "I have not failed, but found 1000 ways to not make a light bulb."

5: Test

Use the steps you created in your numbered list and test to find the problem. Make sure to record your results.

For example, use the steps to find the problem with the lamp. Keep testing until a solution has been reached.

6: Improve

When testing, if you think of any other possible reasons your lamp might not generate light, you can improve the checklist and steps to find a solution to the problem.

The engineering design cycle is a cycle, and by definition cycles are processes meant to be repeated. If at first you can't find the solution, don't give up or lose faith in yourself. Remember to repeat the cycle until you find a solution. That's what makes an engineer an engineer.

For example, you realize that you did not check to see if the switch is working properly.

You can improve your test by adding a step to check if the switch is moving between the on/off positions properly.

Engineering Design Cycle in Practice / Engineering Design Cycle

Each student should choose 1 of the following 3 options. Then use the engineering design cycle to solve the problem.

The three options are:

1. The dishwasher isn't working.

2. The washing machine isn't working.

3. The ice dispenser on the refrigerator isn't working.

What appliance did you choose? ______________________________

Steps in the Design Process	*Student Work*
1. Ask Questions *What questions should you ask?*	

Steps in the Design Process	Student Work
2. Imagine *Think of some possible reasons why your appliance doesn't work.*	

Steps in the Design Process	Student Work
3. Make a Plan *Create a checklist to follow to test the reasons your appliance doesn't work.*	

Steps in the Design Process	*Student Work*
4. Create *Create a numbered list of steps to test the items on your checklist.*	

Steps in the Design Process	*Student Work*
5. Test *Go through the steps with another student or teacher and record the results of the test.*	

Steps in the Design Process	Student Work
6. Improve *After walking through your steps, list any improvements you have thought of or one of your classmates suggests.*	

This process can be used for a variety of other household items. Next time something does not work, don't throw it away, instead use the engineering design cycle to check for solutions first.

ACTIVITY

Paper Table Challenge / Engineering Design Cycle

The paper table challenge is a challenge that requires you to use paper and tape to create a structure that holds up your iPad. The objective is to use the engineering design cycle to come up with a plan to support your iPad with paper.

- The *criteria* is: the table must stably support an iPad for 20 seconds.
- The *constraints* are: only 15 pieces of notebook paper can be used. Only 20 pieces of tape can be used. Your structure should be more than 6 inches tall. Note, you don't have to use all the paper or tape.

1. First ask the question. *What am I trying to accomplish?*

2. Imagine a way to accomplish this goal.

3. Create a plan of exactly how to build the structure. You can use the space on the next page.

4. Build your structure based on your plan.

5. Test your structure but be careful not to drop your iPad.

6. If your structure did not hold, go back, rethink the design and try again, but never give up.

Components	Student Work & Instructions
Materials List	15 sheets of notebook paper 20 pieces of tape Ruler with measurements
Lesson Experiment Following the steps of the engineering design cycle, create your structure. Remember your criteria and constraints. In the space provided below you can draw out your build plans.	

Matching & Review / Engineering Design Cycle

Instructions: Match each term to the correct definition. Please refer back to the previous pages as needed to review this material.

1. **Ask** ______
2. **Imagine** ______
3. **Make a plan** ______
4. **Create** ______
5. **Test** ______
6. **Improve** ______
7. **Constraint** ______
8. **Criteria** ______
9. **Engineering** ______
10. **Checklist** ______
11. **Cycle** ______
12. **Engineering Design Cycle** ______

A. The process used for determining and evaluating solutions to problems.

B. The sixth and final step where the engineer re-evaluates any problems to the checklist and finds ways to improve the checklist.

C. A list of the requirements a solution must meet to be considered successful.

D. The third step where the engineer writes possible solutions into a checklist.

E. The fourth step where the engineer makes a list of steps based on the plan.

F. The list of possible solutions to problems.

G. The fifth step where the engineer tries to solve the problem with the listed steps.

H. The science of problem-solving and design.

I. An iterative, ongoing, and repeatable process.

J. The second step where the engineer brainstorms possible solutions to the problem.

K. The first step where the engineer identifies the nature of the problem.

L. A list of restrictions that limit or control what you can do.

ACTIVITY

Draw the Cycle / Engineering Design Cycle

Extra Time!

In the space below, draw a diagram of the engineering design cycle with pictures and characters to help you remember the steps of the cycle. Use colored pencils and other tools to make it fun and exciting.

THE BROOKLYN BRIDGE / NEW YORK, NEW YORK

Photo by Hannes Richter

LESSON

Engineering Lettering

ROADMAP

- Learn what engineering lettering is.
- Understand why lettering is important.

Engineering Lettering

What is it and why is it important?

Engineering lettering is the process of writing letters and numbers in a way that optimizes **clarity**. Lettering encompasses many different skills such as precision and attention to detail.

Mastering these skills will lead to neater notebooks and project plans. In the classroom, minor mistakes don't usually create big problems and the idea is that you can learn from them. In the real-world minor mistakes can cause catastrophes.

In 1962, one of NASA's rockets, the Mariner 1, had a major catastrophe because of a minor mistake. Due to a software error, the rocket drifted off course.

To avoid the possibility of the rocket falling back to earth, it was blown up. This minor mistake cost around $670 million in 2024 U.S. dollars. This is an example that demonstrates why precision in engineering is so important.

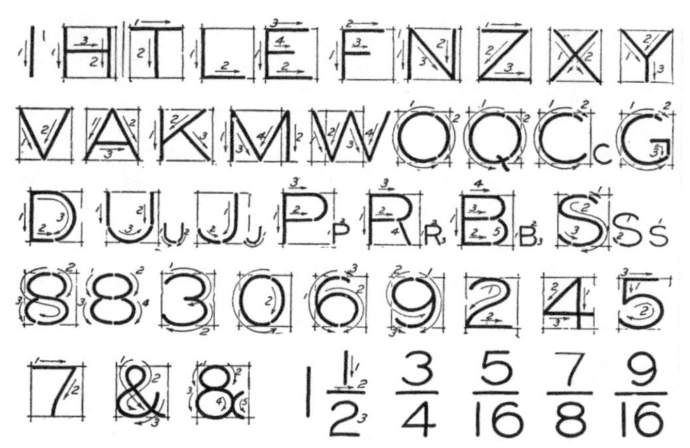

FIGURE 3.1
ENGINEERING LETTERING

In 1999, a mistake was made by NASA that caused a climate orbiter to burn up in Mars' atmosphere. The accident was caused by an engineer who forgot to convert metric units to United States customary units. So check your work!

Vocabulary

Clarity
The quality of being clear or easy to understand.

Single Stroke Letters
In engineering, these are letters written with a line thickness achieved by one stroke of the writing implement.

Neatness
The quality of being visually tidy and orderly.

Consistency
The quality of being uniform in all aspects.

Precision
The quality of being accurate and exact.

Proofreading
The process of reviewing written work to correct errors.

How does it work?

1. Single Stroke Letters:

In an engineering context, **single stroke letters** are letters written with one stroke of the pen or pencil. Although the writing instrument may be lifted from the page when making the letter, the line thickness is one stroke, not two or three.

2. Neatness:

Neatness is very important. Being neat means that the letters are well shaped and easy for anyone to read.

3. Consistency:

Along with neatness, **consistency** is another key factor in making sure mistakes aren't made. When you write a letter, it should look the same every time. Letters must not only be consistent in look, but also all the letters must be consistent in size and spacing. Think about it as if every letter had its own box it needed to fit in.

4. Precision:

Precision is very important when putting dimensions on a drawing. The numbers must be accurate and the lettering must be neat and consistent. To avoid mistakes, the precise location of dimensions on the drawing is also important. Dimensions should never be cluttered together. Being precise means you have taken the time to do things accurately and with detail in mind.

5. Spelling:

All words must be spelled correctly. If you are not certain of a word's spelling, look it up. Never guess the spelling of a word.

6. Proofreading:

Proofreading is extremely important. When you proofread, you double check the legibility and accuracy of written work to avoid mistakes.

Practice Lettering / Engineering Lettering

Using Figure 3.1, follow the steps below to practice the important skill of engineering lettering. Use the graph paper provided on the following pages.

Steps in the Design Process	***Student Work***
Step 1	*Write out every letter in the alphabet and every number three times to ensure you are writing consistently and neatly.*
Step 2	*Once finished, write out the sentence twice:* MY LETTERING MUST BE NEAT AND EVENLY SPACED.
Step 3	*Then write out today's date twice.* Ex: 06/24/1981.
Step 4	*Create your own sentence and practice your lettering.*

Practice Lettering / Engineering Lettering

Practice Lettering / Engineering Lettering

ACTIVITY

Practice Lettering / Engineering Lettering

ACTIVITY

Practice Lettering / Engineering Lettering

COLOSSEUM / ROME, ITALY

Photo by Tommao Wang

Isometric Drawing

ROADMAP

- Learn about the nature and purpose of isometric drawings.
- Learn how to make isometric drawings.

Isometric Drawing

What is it and why is it important?

Today we are going to cover the basics of **isometric drawing** and discover why it is a necessary skill for young engineers to learn.

First, you might be wondering, "what makes something **isometric**?" Isometry means that everything shown is **proportional** to each other no matter how far it is from the viewer. Figure 4.1 shows the difference between what objects look like when they are drawn **three dimensionally** and when they are drawn isometrically. The top section of Figure 4.1 is drawn three dimensionally and the bottom is drawn isometrically.

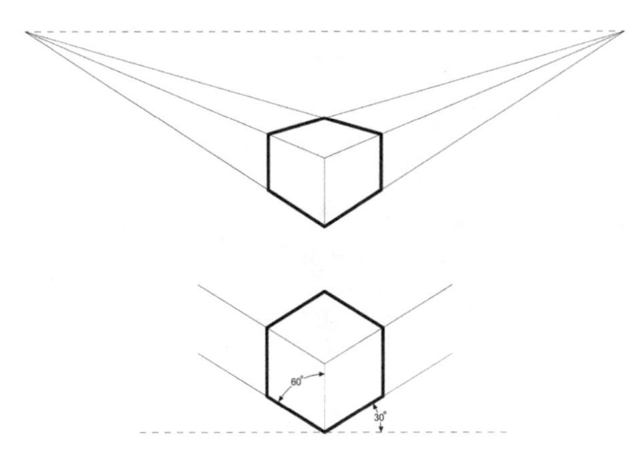

FIGURE 4.1
3D & ISOMETRIC PERSPECTIVE

In a three dimensional view, objects are distorted by perspective. As objects get farther away from the viewer, they will appear smaller than they truely are. Isometric drawings allow you to see the correct size of all three dimensions of an object in a single view. This means that engineers can see exactly what each side of an object should look like without being distorted by **perspective**. The length, width, and height all remain correctly sized to the viewer no matter how far off in the distance the object is shown.

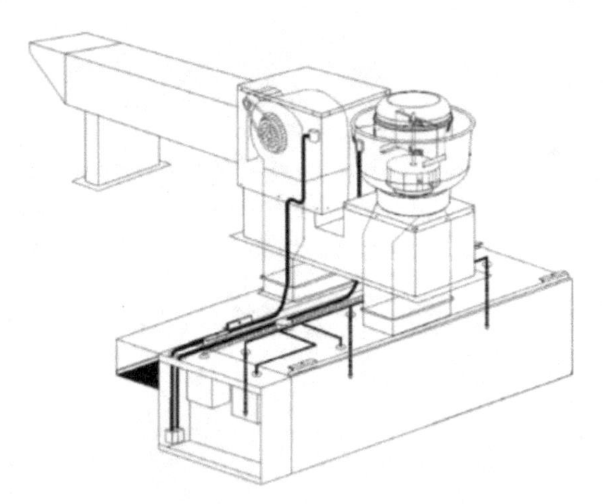

FIGURE 4.2
ISOMETRIC DRAWING OF A COMMERCIAL KITCHEN HOOD

Now you might be wondering, "how is something as simple as isometric drawing going to help me with my engineering skills?" Look at everything around you: the building you are in, the desk you are sitting at, the chair you are sitting in. All these things were built because someone used isometric drawing to plan it out beforehand. Figure 4.2 is a picture of an isometric drawing of a commercial oven hood that an engineer has used to help build it.

Although we will not be drawing anything this complicated today, it is interesting to see what these skills can be used for.

Vocabulary

Isometric Drawing
A method of representing a three-dimensional object at full scale by aligning it with the X, Y, and Z axes.

Isometry
A method of representing a three-dimensional object at full scale by aligning it with the X, Y, and Z axes.

Proportional
When two different sized objects are equally scaled regardless of place within a larger arrangement.

Three-Dimensional
An object that has length, width and height.

Perspective
A view of a three-dimensional object which gives it the impression of depth.

Axis
An imaginary line through an object used to display the object's size and location.

How does it work?

When drawing isometrically, it is important to remember that you are not just drawing a picture of an object, but planning out what the object would look like in real life.

In order to do this, you must remember some important things:

1. Always make sure that each line you draw stays parallel to its axis. There are three axes in isometric drawing: X, Y, and Z. These three imaginary lines correspond to the length, width, and height of the object. The X, Y, and Z axes all meet in the center in the shape of a "Y" and are perpendicular to each other. Figure 4.3 is an isometric projection of a cube. The "Y" is drawn in red so that you can see where the X, Y, and Z axes come together.

2. Aways draw neatly using a sharp pencil. Make sure to draw lightly so that if you need to erase something, your diagram will still be neat.

3. *Always Use a Ruler!* Not using a ruler is one of the easiest ways to be inaccurate with your measurements and ruin your diagram.

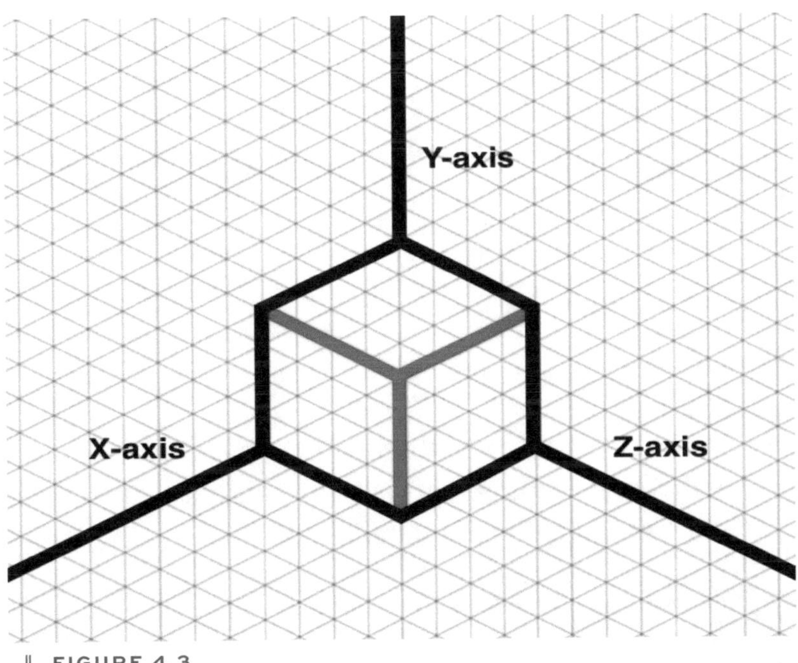

FIGURE 4.3
ISOMETRIC CUBE

Now that you have the basics for isometric drawing down, we can practice drawing some isometric projections.

Note that all of your lines are drawn either horizontally or at a 30 degree angle. Isometric paper has the lines drawn horizontally and at 30 degree angles to help you.

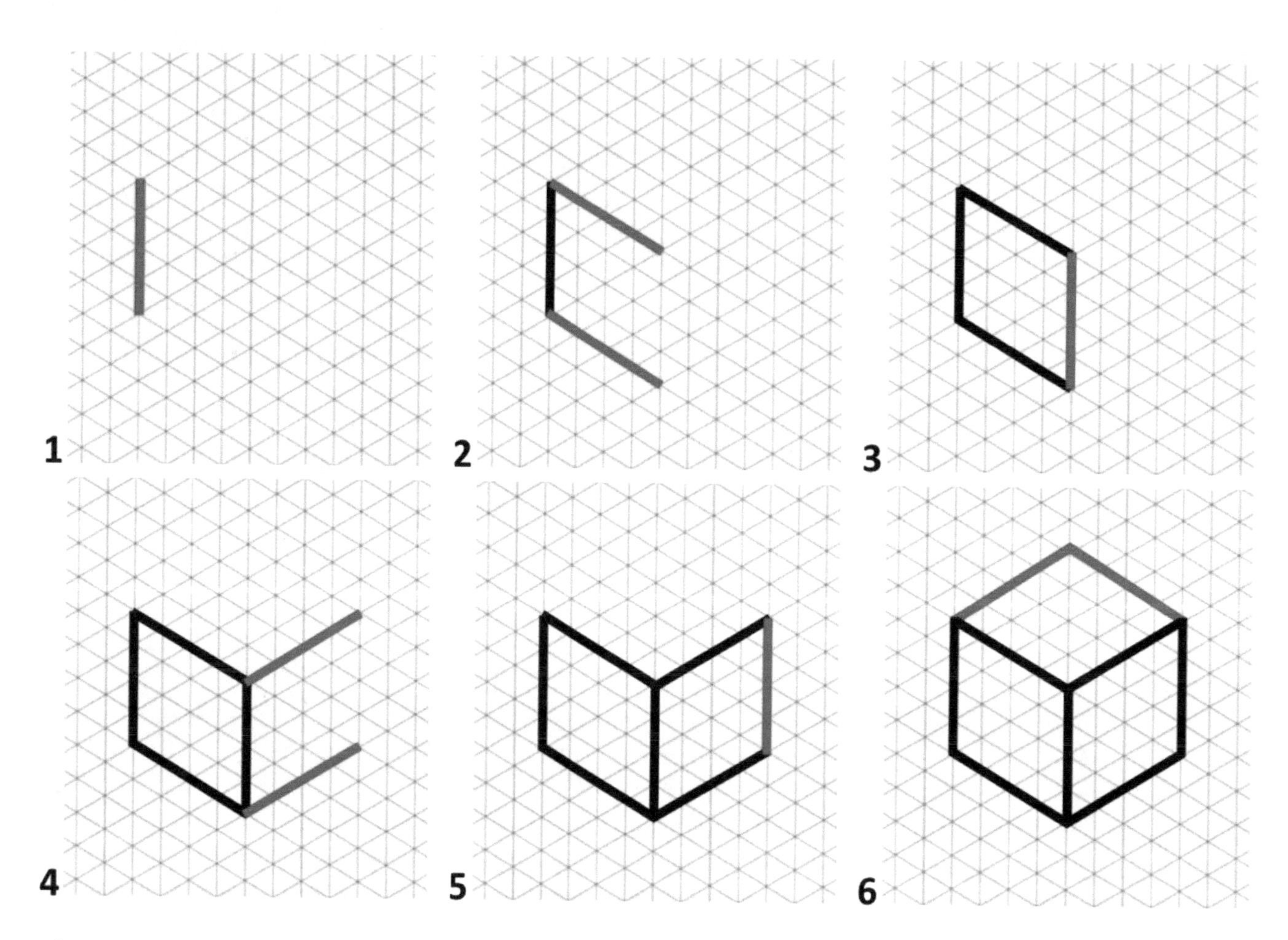

FIGURE 4.4
STEPS TO DRAWING AN ISOMETRIC CUBE

Your lines do not have to be three units long. You can draw them using only one or two units if you like. However, because this is a cube and has equal sides, you must keep them proportional. All of your lines must be the same length.

As you do your in-class exercise, refer to Figure 4.4 and follow the steps to draw perfect isometric cubes.

Practice Drawing / Isometric Drawing

Let us practice the skills of isometric drawing with the following materials and resources. For this assignment, you will need:

- Sharpened pencil
- Ruler with measurements
- 10 colorful cubes

Step 1	Instructions
FIGURE 4.5	Take out two cubes, attach them together, and place them on the table in front of you. (Refer to Figure 4.5) Use your isometric paper to draw a three-dimensional projection of the cubes. Make sure your lines are parallel! Once you have completed your first isometric projection, wait for the teacher to check your work and then move on to step 2.
Step 2	**Instructions**
FIGURE 4.6	Take out three more cubes and attach them to the original two cubes. (Refer to Figure 4.6) Now, turn the 5-block construction slightly diagonal and draw an isometric projection of it. Once you are finished, wait for a teacher to check your work and then move on to step 3.

Practice Drawing / Isometric Drawing

Step 3	Instructions
 FIGURE 4.7	Take out two more cubes and attach them together in the shape of the letter H. (Refer to Figure 4.7) Stand it up on your desk, turn it diagonal, and draw an isometric projection of your construction. Once you are finished, wait for a teacher to check your work and then move on to step 4.

Step 4	Instructions
 FIGURE 4.8	Take out the remaining three cubes so that all ten cubes are on your desk. Attach the ten cubes together in any way you like and place the assembly flat on the table. An example of what this might look like is shown in Figure 4.8. You must build a construction that is different from the one shown in Figure 4.8. Draw an isometric projection of your creation, making sure to keep your measurements proportionate and your lines straight.

The beginnings of isometric drawing date all the way back to the ancient Chinese who used a form of perspective called axonometry to give art a three-dimensional appearance. Before the beginning of axonometry over 2,000 years ago, ancient art was only two-dimensional. However, by using axonometry, Chinese artists discovered how to draw three-dimensionally using parallel lines on a single plane. This advancement changed art forever and was one of the first steps toward the more modern art form known as Realism. Figure 4.9 is an image which shows one of the earliest uses of axonometry.

ACTIVITY

Practice Drawing / Isometric Drawing

ACTIVITY

Practice Drawing / Isometric Drawing

ACTIVITY

Practice Drawing / Isometric Drawing

LONDON BRIDGE / LONDON, UNITED KINGDOM

Photo by Bush 'o' Graphy

CHAPTER

Orthographic Drawing

ROADMAP

- Learn about the nature and purpose of orthographic drawings.
- Learn how to make orthographic drawings.

Orthographic Drawing

What is it and why is it important?

Now that you have explored isometric drawing down, it is time to learn a similar skill which is very useful for all types of engineering design: **orthographic drawing**.

Unlike isometric drawings, orthographic projections do not give a full, three-dimensional view of an object. Instead, orthographic projections provide engineers with at least three **two-dimensional** views for each side of an object so that the measurements of every individual component can be easily seen. This is an important tool for engineers since it is easier to communicate the specific details of ideas using two-dimensional projections, while three-dimensional projections are good at showing the final product. In Figure 5.1, an isometric drawing of a simple object is split into orthographic projections for its top view, front view, and side view.

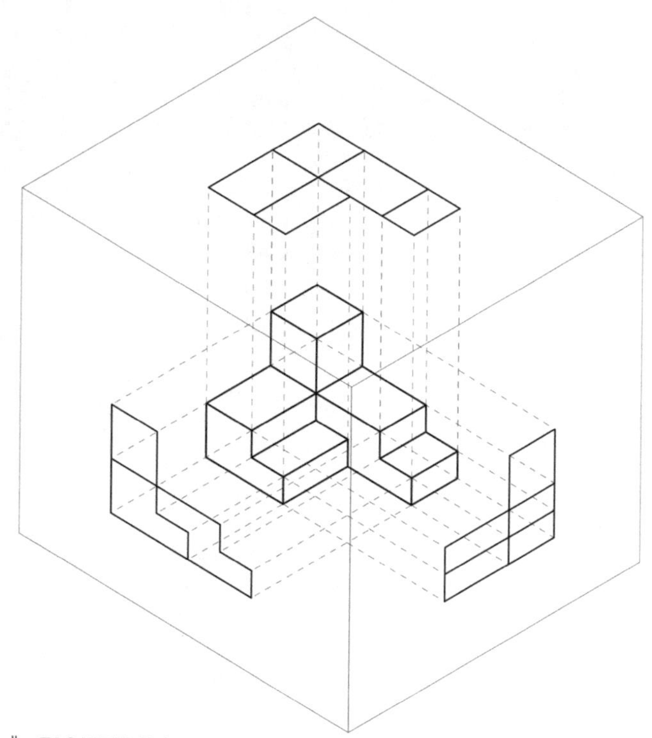

FIGURE 5.1
AN ORTHOGRAPHIC PROJECTION

Orthographic projections have been used throughout most of history, but one man is most famous for using them. During the first century BC, a Roman architect and engineer named Marcus Vitruvius Pollio, who is pictured on the far right presenting a collection of his ideas called De Architectura *to Augustus, used orthographic projections to design his first functioning sundial.*

Vitruvius himself gave orthographic drawing its name. The word orthographic *comes from the Greek word "orthos," which means* perpendicular *or* straight.

Vocabulary

Orthographic Drawing
A method of representing a three-dimensional object in two dimensions from at least three two-dimensional sides.

Two-Dimensional
An object that has length and width.

How Does it Work?

An orthographic projection is a way of representing a three-dimensional object in two dimensions. Orthographic projections show exactly what one side of the object would look like. They include everything visible from that perspective as well as all the measurements. Most orthographic projections only require three views to represent the object: the front view, the top view, and the right side view, but sometimes more are required.

Usually, engineers use only three projections: the front, the top, and the right side. However, if there is a feature (like a hole) that is not visible using the three typical projections, you would need to add an additional projection to show that feature. We can see this in Figure 5.1. The back of the house does not have windows like the front of the house does, so more orthographic projections are needed to show the features that are different on the back of the house.

Using this system of orthographic projections, you can easily see how the projections connect together to form the isometric drawing. Figure 5.2 is a diagram of some orthographic projections which use these three perspectives.

Whenever you draw out three orthographic projections, make sure to arrange your projections in the order that you see them in Figure 5.2.

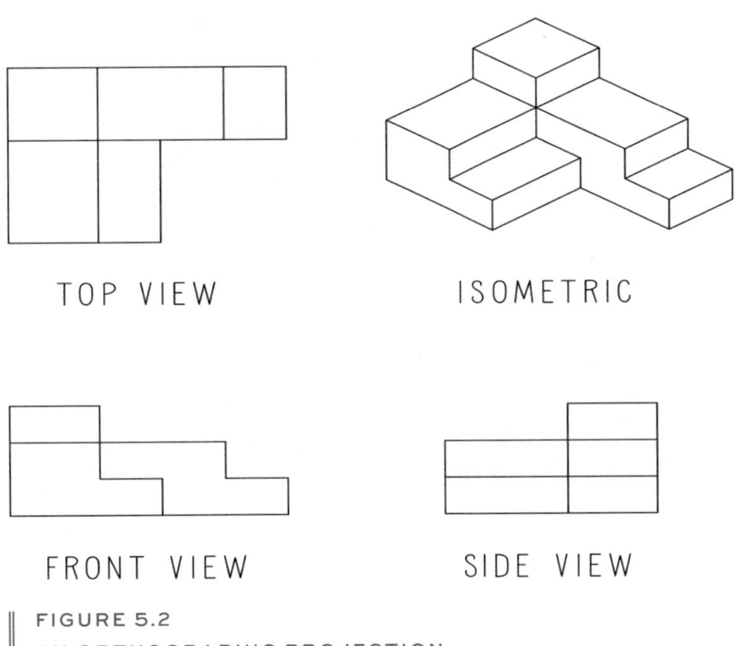

FIGURE 5.2
AN ORTHOGRAPHIC PROJECTION

ACTIVITY

Practice Drawing / Orthographic Drawing

Let us practice the skills of orthographic drawing with the following materials and resources. For this assignment you will need:

- Sharpened pencil
- Ruler with measurements
- 10 colorful cubes

Before starting, remember that you are drawing cubes. Cubes have the same length, width and height. Since the length, width and height are equal, all of the lines you draw must be the same length. You may use any length you want for each line as long as the lines for the length, width and height are all the same.

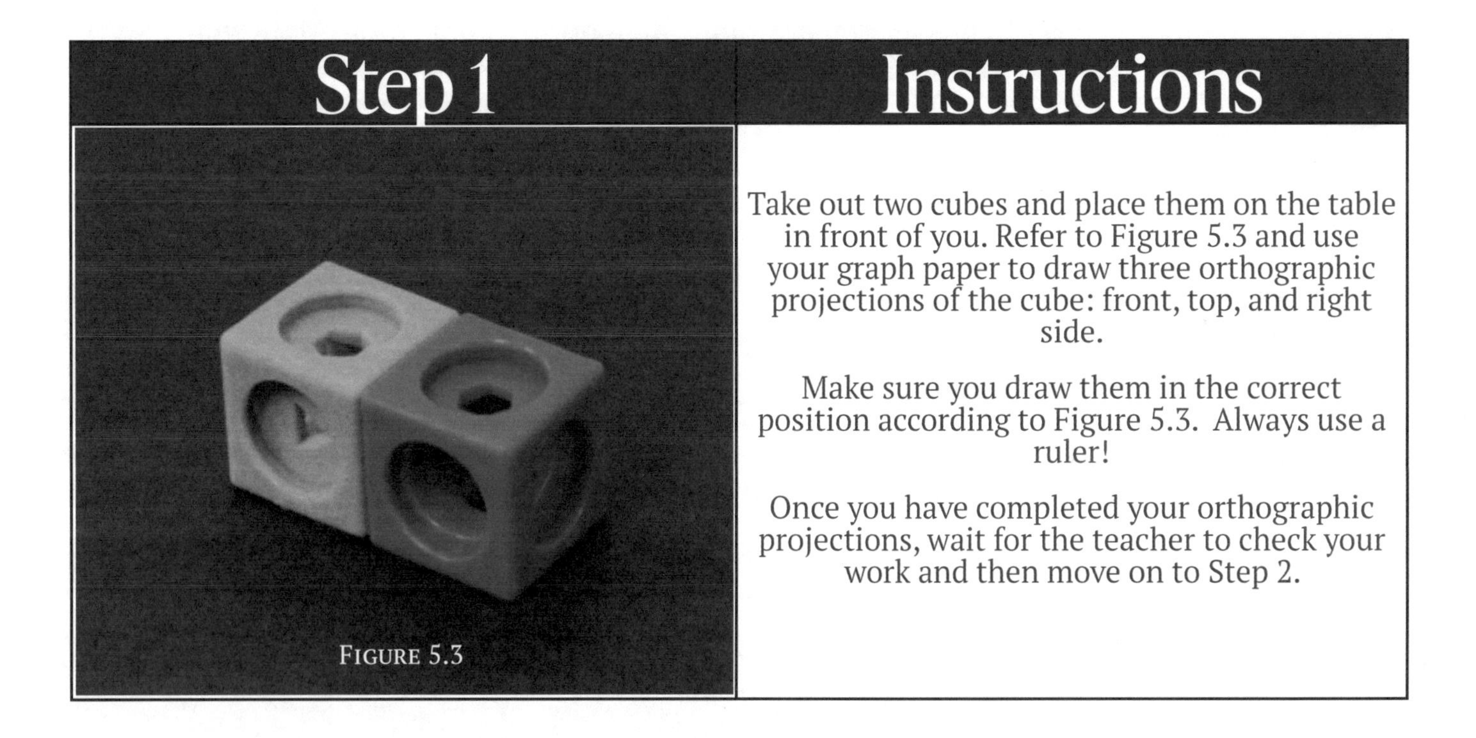

Step 1	Instructions
FIGURE 5.3	Take out two cubes and place them on the table in front of you. Refer to Figure 5.3 and use your graph paper to draw three orthographic projections of the cube: front, top, and right side. Make sure you draw them in the correct position according to Figure 5.3. Always use a ruler! Once you have completed your orthographic projections, wait for the teacher to check your work and then move on to Step 2.

ACTIVITY

Practice Drawing / Orthographic Drawing

Practice Drawing / Orthographic Drawing

Continue the assignment on *orthographic drawing* you began on the previous page.

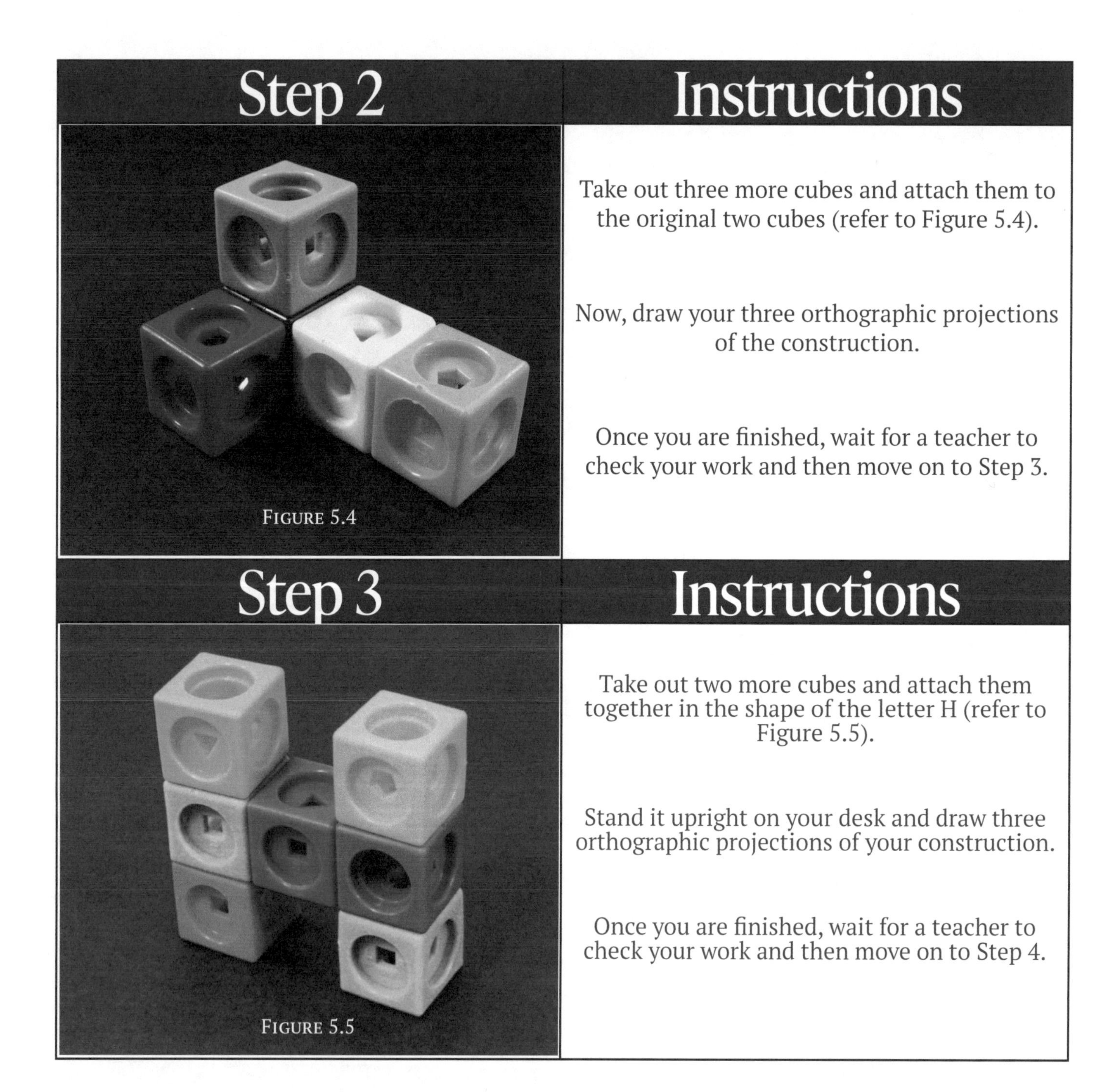

Step 2	Instructions
FIGURE 5.4	Take out three more cubes and attach them to the original two cubes (refer to Figure 5.4). Now, draw your three orthographic projections of the construction. Once you are finished, wait for a teacher to check your work and then move on to Step 3.
Step 3	**Instructions**
FIGURE 5.5	Take out two more cubes and attach them together in the shape of the letter H (refer to Figure 5.5). Stand it upright on your desk and draw three orthographic projections of your construction. Once you are finished, wait for a teacher to check your work and then move on to Step 4.

Practice Drawing / Orthographic Drawing

Continue the assignment on *orthographic drawing* you began on the previous page.

Step 4

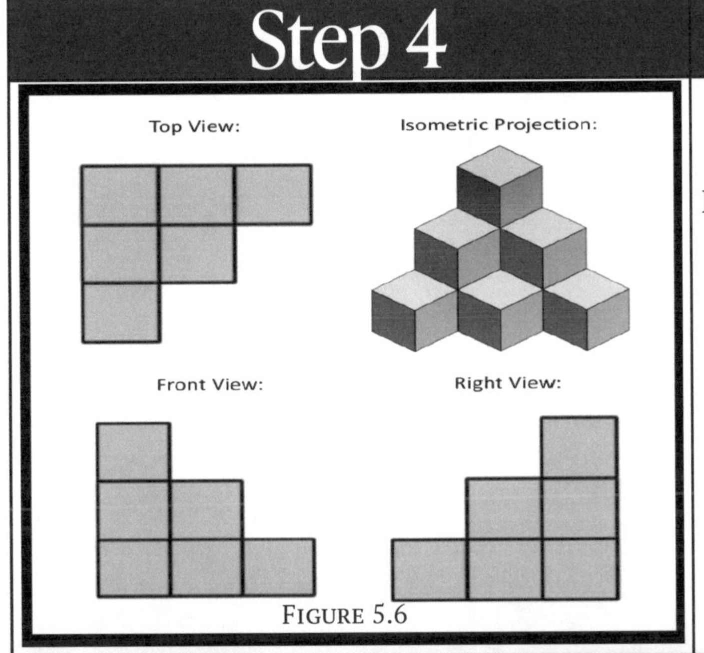

FIGURE 5.6

Instructions

Look back to your isometric projection of the 10-block construction from Step 4 of Lesson 4.

Now, without putting your cubes together, draw the three orthographic projections (front, top, and right side) of your 10-block construction.

You may only use your isometric drawing of the 10-block construction for this step. Figure 5.6 is an example of what your orthographic projections could look like, but you should draw your projections according to your own construction.

Once you are finished, wait for a teacher to check your work and then move on to Step 5.

Step 5

FIGURE 5.7

Instructions

Take out the remaining three cubes so that all ten cubes are on your desk.

Using only your orthographic projections, reconstruct the same 10-block construction that you did last lesson. Figure 5.7 is an example of what the construction from Step 4 would look like if you built it according to the projections.

Wait for your teacher to check your work and make sure the isometric and orthographic drawings match up correctly.

Practice Drawing / Orthographic Drawing

Practice Drawing / Orthographic Drawing

TREVI FOUNATIN / ROME, ITALY

Photo by Cristina Gottardi

6

LESSON

Metrology

How to Measure with a Ruler, Caliper, Tape Measure & Speed Square

ROADMAP

- Learn about metrology, which is the *science of measurement.*
- Learn how to use a ruler, caliper, tape measure, and speed square.
- Know how to use the least count to read markings on measuring instruments.
- Know how to measure length using a ruler.

Metrology: How to Measure with a Ruler, Caliper, Tape Measure, & Speed Square

What is it and why is it important?

One of the most important skills a person can have is understanding how to measure things. Without accurate measurements, engineers would not be able to build anything because they would not know what size to make each piece.

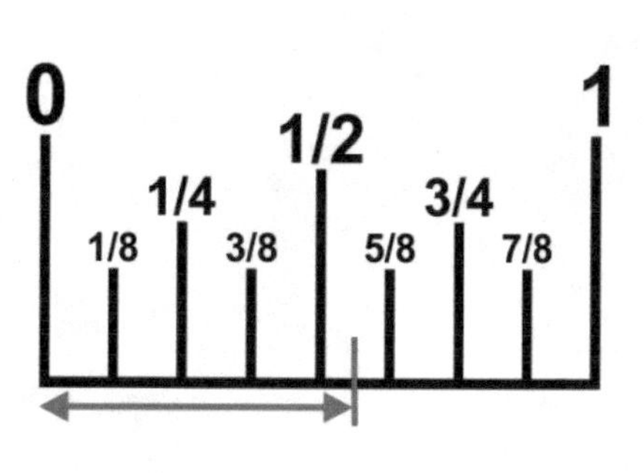

FIGURE 6.1
LEAST COUNT

Understanding how to measure different lengths and mark them with different tools is a very useful skill. Today, we will learn to use four different tools that will help you measure accurately: ruler, caliper, tape measure, and speed square.

It is important to remember that when you measure anything, you must only round your measurement to one-half of the smallest length marked, also called the **least count** of the measuring instrument. This means that if the smallest measurement on your ruler is ⅛ in, then the most precise measurement you can record is down to 1/16 in.

Figure 6.1 is a diagram of one inch that has been divided into sections ⅛ in long. If you measure the red arrow, it is too long to measure ½ and too short to measure ⅝ in. In this case you would estimate the red arrow to be 9/16 in long, which is directly between the two. Do not attempt to be more precise than one half of the smallest measurement shown.

How Does it Work?

Tool #1: Rulers

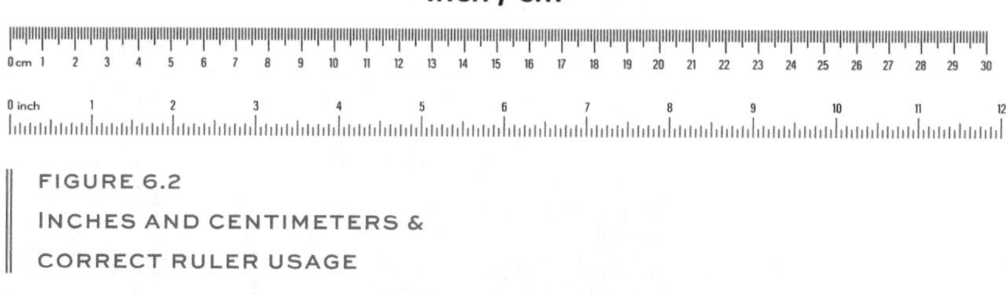

FIGURE 6.2
INCHES AND CENTIMETERS & CORRECT RULER USAGE

Rulers are the most common measuring devices used by engineers for drawing. Using a ruler is one of the easiest ways to draw straight lines with accurate measurements. Most 12 inch rulers have inches on one side and centimeters on the other, so whichever unit you use is up to you. See Figure 6.2 for an example. Whenever you measure with a ruler, follow these simple steps.

Vocabulary

Least Count
The smallest measurement that can be accurately determined from a measuring instrument; It is 1/2 of the smallest unit on the measuring device.

Ruler
A straight edge tool used to measure length and draw straight lines

Caliper
A measuring instrument with two clamps used to accurately determine the length of an object or the distance between two objects.

Tape Measure
A flexible measuring instrument that is used to accurately determine the length of larger objects and is able to roll into a compact case

Speed Square
A triangular, multipurpose tool used for measuring, drawing straight lines, and determining square angles

1. Set the 0 inch marker (or centimeters, whichever you are using) exactly where you want to begin measuring. Usually this is the edge of the object you are measuring.
2. Make sure to keep the ruler on a flat surface so that your measurement remains accurate.
3. Look at the end of the object you are measuring and determine how long it is using the markings on the ruler.

Tool #2: Calipers

Calipers are primarily used for two different things. The first is measuring length by placing the two larger clamps on either side of the object and reading the ruler to see how long it is, as shown in Figure 6.3.

The second is measuring the distance between two objects by placing the two smaller clamps on the edge of each object and reading the ruler to see the distance between the two, as shown in Figure 6.4. Calipers are most commonly used for getting precise measurements of small objects.

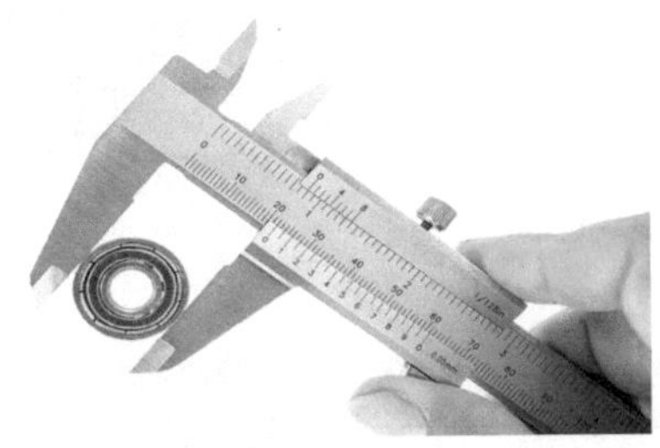

FIGURE 6.3
OUTSIDE MEASURE

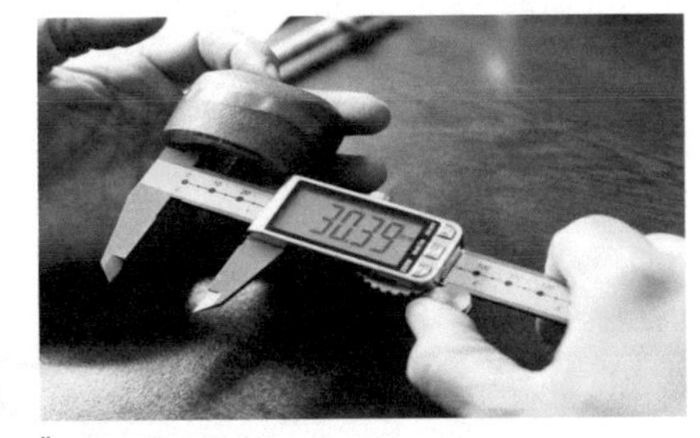

FIGURE 6.4
INSIDE MEASURE

Note: If you are using an electronic caliper as shown in Figure 6.4, make sure to zero out the reading before measuring. Whenever you press the ZERO button, the reading on the caliper display will be set to zero. Make sure to zero out the caliper before you measure an object, otherwise your reading will be incorrect.

Tool #3: Tape Measure

A **tape measure** is one of the most common tools used in engineering. Although it is not used for drawing straight lines, a tape measure is able to give you the measurements of larger objects. This is because a tape measure is much longer than a ruler. An example of a tape measure is shown in Figure 6.5.

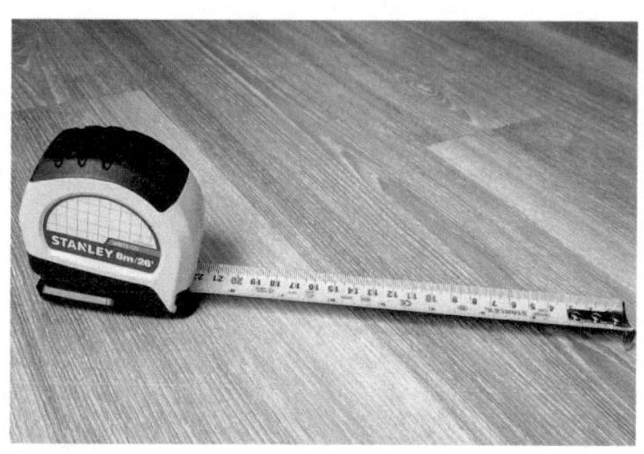

FIGURE 6.5
TAPE MEASURE

Most tape measures are anywhere from 15 to 50 feet long. A tape measure is typically more consistent over long distances than a ruler, so if you need to measure something large, a tape measure is your best bet.

Tool #4: Speed Square

A **speed square**, as shown in Figure 6.6, can be used for many things, but it is especially useful when working with wood. Speed squares are perfect for drawing straight, perpendicular lines across a piece of wood. In order to do this, you must line up the lip of the speed square with the edge of your wood and use the ruler side to draw a straight line across your board, which is shown in Figure 6.7.

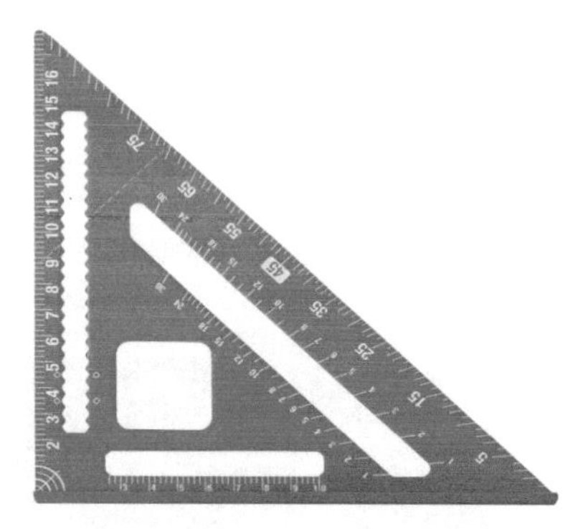

FIGURE 6.6
SPEED SQUARE

FIGURE 6.7
A SPEED SQUARE IN USE

The first speed square was invented by a carpenter named Albert J. Swanson in 1925. Swanson initially created the tool in order to assist him with measuring roof pitches. As demand for the tool grew, he began mass-producing them. He founded the Swanson Tool Company in 1945 and it gradually expanded to produce and distribute a wider variety of carpentry tools. Today, the Swanson Tool Company is based in Frankfort, Illinois and is one of the largest tool companies in America.

Measuring Practice / Metrology

Let us practice the skills of measuring with tools using the following materials and resources. For this assignment, you will need:

- Group of 2 or 3 students
- Sharpened pencil
- Ruler (Per group of 2 or 3)
- Caliper (Per group of 2 or 3)
- Tape measure (Per group of 2 or 3)
- Miscellaneous school supplies

Step	Instructions
Step 1	Have each student in your group take one of the three tools in front of you (ruler, caliper, and tape measure). On your desk should be some everyday school supplies. Choose one of them and use whichever tool is suitable to measure the length of something on the desk or in the room. Then record it in your workbook on the pages provided. Some examples of measurements are listed below. Ruler: length of light switch cover Caliper: length of an Expo marker Tape measure: width of the classroom door
Step 2	When you have finished measuring your first object, write down in your workbook what you measured, what you used to measure it, and how long it was. Here is an example of what you might write down: *Caliper, Expo marker: 5 in*

Measuring Practice / Metrology

Step	Instructions
Step 3	Once the other students in your group have finished writing down their measurements, swap tools and choose another item to measure using your new tool. Just as before, write down what you measured, what you used to measure it, and how long it was. Then swap tools again and do the same for the third tool.
Step 4	When all students in your group have one object measured for all three tools, swap workbooks with someone in your group so you now have their workbook with their measurements. Look at each of the three objects your partner measured and use the same tools they used to double check their answers. Put a check mark next to his or her measurements if they are accurate. Return the corrected workbook to your partner.

ACTIVITY

Measuring Practice / Metrology

DURHAM CATHEDRAL / DURHAM, ENGLAND

Photo by Charles Givens

LESSON

Metrology

Measuring with Different Unit Systems

ROADMAP

- Understand the difference between U.S. customary and metric units.
- Learn how to measure length with a ruler in both U.S. customary and metric units.
- Learn how to measure mass with a digital scale in both U.S. customary and metric units.
- Know what the TARE button does on a digital scale.
- Learn how to measure temperature with a thermometer in both U.S. customary and metric units.
- Be able to find the length, mass, or temperature of an object in the correct units.

Metrology: Measuring with Different Unit Systems

What is it and why is it important?

Did you know that the United States is one of the only countries in the world that measures in inches? The unit system commonly used in the U.S. is called the **U.S. Customary System**, which orriginated from the British Imperial System that was used in British colonies. However, the majority of countries globally use the **International System of Units** (SI) and is more commonly known as the **metric system**, which was invented in France during the French Revolution in the late 18th century to standardize the various unit systems used throughout France .

Almost every tool you use in engineering will have both metric and U.S. customary unit markings. Last lesson, you measured the lengths of different objects using inches. Today, we will be measuring not only length, but also mass and temperature, and we will be using both the U.S. customary system and the metric system.

How Does it Work?

1. Length

As you already know, **length** is the measurement of how long an object is. In the last lesson, you learned to measure using inches (in), but most of the world uses the metric unit which is centimeters (cm). Centimeters are much smaller than inches as you can see in Figure 7.1.

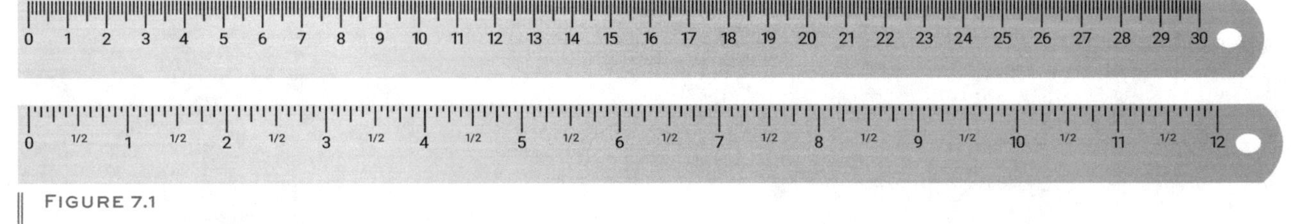

FIGURE 7.1
INCHES & CENTIMETERS

2. Mass

Mass is the measure of how much matter, or substance, is in an object and will not change unless you add more matter to the object. The U.S. customary system uses ounces (oz) to measure the mass of objects while the metric system uses grams (g). Figure 7.2 is an example of what a digital mass scale looks like.

FIGURE 7.2
OUNCES & GRAMS

Objects feel heavy because of their **weight**, which is how heavy an object is due to its mass under the effect of gravity, since you have to resist gravity to lift them.

A·Z

Vocabulary

U.S. customary system
The system of weights and measurements most commonly used in the United States.

International System of Units
The system of weights and measurements most commonly used by the rest of the world and also called the Metric System.

Metric System
Another name for the international system of units.

Length
The measurement of how long or short an object is in inches or centimeters.

Mass
The measurement of how much matter is in an object in ounces or grams.

Weight
The measurement of how heavy an object is based on the gravitational pull it is under.

Tare
To adjust a scale so as to reduce the displayed weight to zero.

Temperature
The measurement of how hot or cold an object is in Fahrenheit or Celsius.

Mercury
The only metal that is liquid at room temperature; historically used in analog thermometers.

To turn the scale on, press the ON/OFF button. Then, make sure that your units are set to the unit system you wish to use (ounces or grams). To switch between ounces and grams, press the MODE button. You may also notice a button that says **Tare**. Whenever you press the TARE button, the reading on the scale will be set to zero. This is also called zeroing out the scale. Make sure to zero out the scale before measuring the mass of an object, otherwise your reading will be incorrect. Then, place your object on the scale and determine both the reading and the units as shown on the digital display.

3. Temperature

Temperature is the measure of the amount of heat present in an object. The U.S. customary measurement for temperature is degrees Fahrenheit (°F), and the metric measurement is degrees Celsius (°C). Some thermometers are analog and use a dial to show you the temperature, while others are liquid thermometers and may use a fluid called **mercury** to show you the temperature.

You can determine the temperature in either Fahrenheit or Celsius by reading how far up the thermometer's scale the liquid reaches. Figure 7.3 is an image of an analog thermometer. Figure 7.4 is another example of a liquid thermometer which reads around 42°C or 108°F.

If you are using a digital thermometer, read the number and unit shown on the digital display. Using a digital thermometer is very similar to using a digital scale. To turn the thermometer on, press the ON/OFF button. Then, make sure that your units are set to the correct units you wish to use (Fahrenheit or Celsius). To switch between Fahrenheit and Celsius, press the MODE button. Figure 7.5 is a picture of a digital thermometer.

FIGURE 7.3
ANALOG THERMOMETER

Now that you understand the basics of how these tools work, let us practice using them with a few activities.

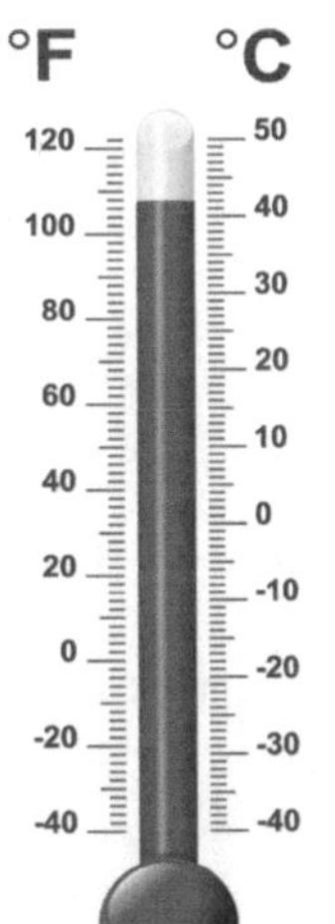

FIGURE 7.4
LIQUID THERMOMETER

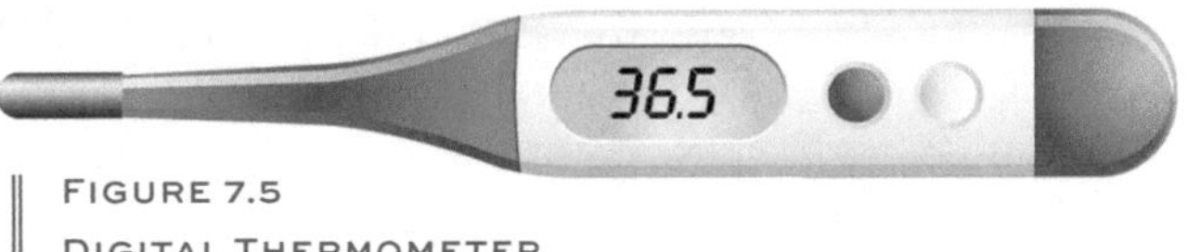

FIGURE 7.5
DIGITAL THERMOMETER

Practice Measuring / Metrology

There will be one of each tool and two labeled cups of sand for every group provided by the teacher. In order to find the temperature of the sand in a cup, push the bottom of the soil thermometer into the sand and determine the reading. The two cups of sand are ONLY for measuring temperature. For this assignment, you will need the following items for every group of 2 to 3:

- Tape measure with inches and centimeters
- Digital scale with ounces and grams
- Soil thermometer with Fahrenheit and Celsius
- Liquid thermometer with Fahrenheit and Celsius
- Cup of cold clean sand
- Cup of room temperature clean sand

Step	Instructions
Note	There should be one of each tool for every group. There will also be two labeled cups of sand provided by the teacher. In order to find the temperature of the sand in a cup, push the bottom of the thermometer into the sand and determine the reading. The two cups of sand are ONLY for measuring temperature.
Step 1	Use the tape measure, scale, and thermometer mentioned above to answer the following questions. Make sure to include the correct units in your answers. Here is a word bank of the possible units your answers could include: in, cm, oz, °F, °C When you are finished, compare your measurements with a friend's measurements.

Practice Measuring / Metrology

1. How long is your desk in inches? ________

2. What is the mass of your pencil in grams? ________

3. What is the temperature of the cup of cold sand in Fahrenheit? ________

4. How long is your notebook in centimeters? ________

5. What is the mass of your notebook in ounces? ________

6. What is the temperature of the cup of cold sand in Celcius? ________

7. How wide is the classroom door in inches? ________

8. What is the mass of the tape measure in grams? ________

9. What is the temperature of the cup of room temperature sand in Fahrenheit? ________

10. How long is your pencil in centimeters? ________

11. What is the mass of the tape measure in ounces? ________

12. What is the temperature of your classroom in Celsius? ________

Crossword Puzzle / Metrology

Across	Down
2. What is the U.S. customary unit for measuring temperature?	1. What is a U.S. customary unit for measuring mass?
4. What is the 80th element on the periodic table, which is a heavy silvery-white metal and a liquid at room temperature?	3. What is it called when you set the digital display on an electric measuring instrument to zero?
7. What is a metric unit for measuring length?	5. What is the metric unit for measuring temperature?
8. What is a metric unit for measuring mass?	6. What is the unit system that most of the world uses?
9. What is a U.S. customary unit for measuring length?	

ACTIVITY

Create a Word Search / Metrology

If you have free time, try to create a word search with the words from your crossword puzzle. Then exchange workbooks with a friend and try to solve each other's word search.

The first working scale was invented by the Ancient Egyptians over 5,000 years ago. Although these scales did not display the exact mass of the object, the Ancient Egyptians were able to use them to determine which objects had more mass than others. These scales were called balance scales and can be seen in many pieces of Egyptian artwork. Balance scales were made of one vertical beam with two pans hanging from a cross beam for massing the objects.

The Egyptians would place weights on one side of the scale and the object they wished to mass on the other side. The scale would tip towards the heavier object and would remain balanced if they had the same mass. The painting to the left is of Anubis, the Egyptian god of the dead, using a balance scale.

GRAND CENTRAL STATION / NEW YORK, NEW YORK

Photo by Pavol Svantner

8 & 9

LESSON

Packaging

ROADMAP

- Learn about why packaging is important.
- Learn how to build a strong and precise product package.
- Understand the process of building a package.

Packaging

What is it and why is it important?

Today we will be going over the basics of packaging. As transportation technology has improved, the long distance transit of objects has increased significantly. To prevent damage to these objects while they were being transported, it became effective to encase them in a low-cost and durable material. Packages must be designed with the needs of their specific contents in mind, while taking up as little space as possible. They need to fit as closely around the contents as possible to prevent the object inside from being damaged and reduce costs, while allowing easy access to the contents. Figure 8.1 depicts a package that fills all the requirements of a proper package.

FIGURE 8.1
COMPLETED BOX

The package in Figure 8.1 has a close fit around the internal components, a strong top and overall construction to keep the objects protected.

Did you know that the first paper commercial board box was used in 1817 in Germany for the packaging of a board game? The game is shown here. In the same year, another person named Malcom Thornhill was credited by many sources who said he made a paper box, but it is actually a mystery whether he actually existed.

Vocabulary

Graph Paper
Paper printed with small squares to assist when drawing graphs or other diagrams

Cardstock Paper
A type of paperboard that is thicker and more durable than normal writing and printing paper, but thinner and more flexible than other forms of paperboard

Scoring Board and Scoring Tool
These tools are used for the process of scoring the paper to make the fold easier. A scoring board and tool are used by putting the line for a fold over a grove in the board and then using the scoring tool to run down that line creating a crease.

How Does it Work?

When creating a package, you must remember that space is key. The package should only just be big enough to protect the contents and that's all. To create an effective package, you must do these seven things.

1. Find the dimensions of the object you are trying to package. Find the width, height, and length. Be sure to keep your units the same. (If using inches, only use inches, if using centimeters, only use centimeters)

2. Remember to leave some extra space for your item. It's a good rule to give yourself about an extra 1/4 to 1/8 inch or ¾ to ⅓ centimeter of clearance on each side.

3. Remember to leave glue tabs on each side of the package. Generally, leave anywhere from 1 inch to ½ inch (2.5 to 1.3 centimeters) of tab on the side to ensure a strong contact area. Figure 8.2 shows an image of a glue tab.

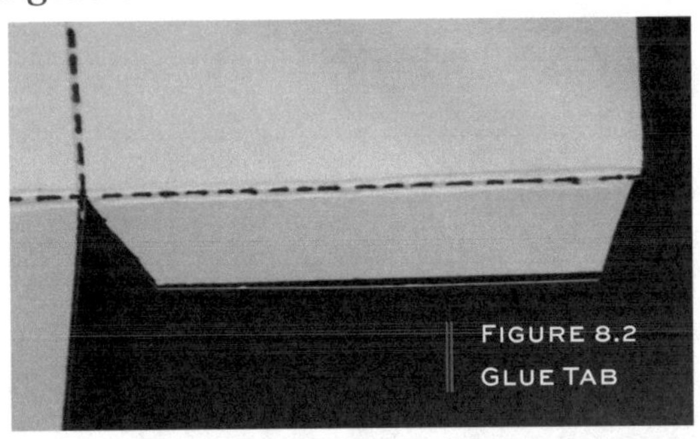
FIGURE 8.2
GLUE TAB

4. Use a ruler or straight edge for drawing every line on the **graph paper** or the **cardstock paper.** If one line isn't straight or the right size it changes the size and the shape of the box and the contents might not fit properly.

5. Always use a **scoring board and a scoring tool** to make folds. It is more accurate and creates a more even fold.

6. Dotted lines represent lines where folds should be.

7. Solid lines represent lines that should be cut.

Package Design / Packaging

Let us practice the skills of *packaging* by designing and making a package with the following materials and resources. For this assignment, you will need:

- Cardstock paper
- Sharpened pencil
- Ruler with measurements
- Scoring board and scoring tool: Figure 8.13
- Glue gun
- Scissors
- Ball
- Graph Paper

Step 1	Instructions
 Figure 8.3	Measure the length, width, and height of your ball (all three dimensions will be the same since it is a sphere, but with other shapes the dimensions will vary). Write these dimensions down in the table below, both in inches and in centimeters.

Student Work

Length: ______________________________

Width : ______________________________

Height: ______________________________

Package Design / Packaging

Instructions: Continue the exercise from the previous page.

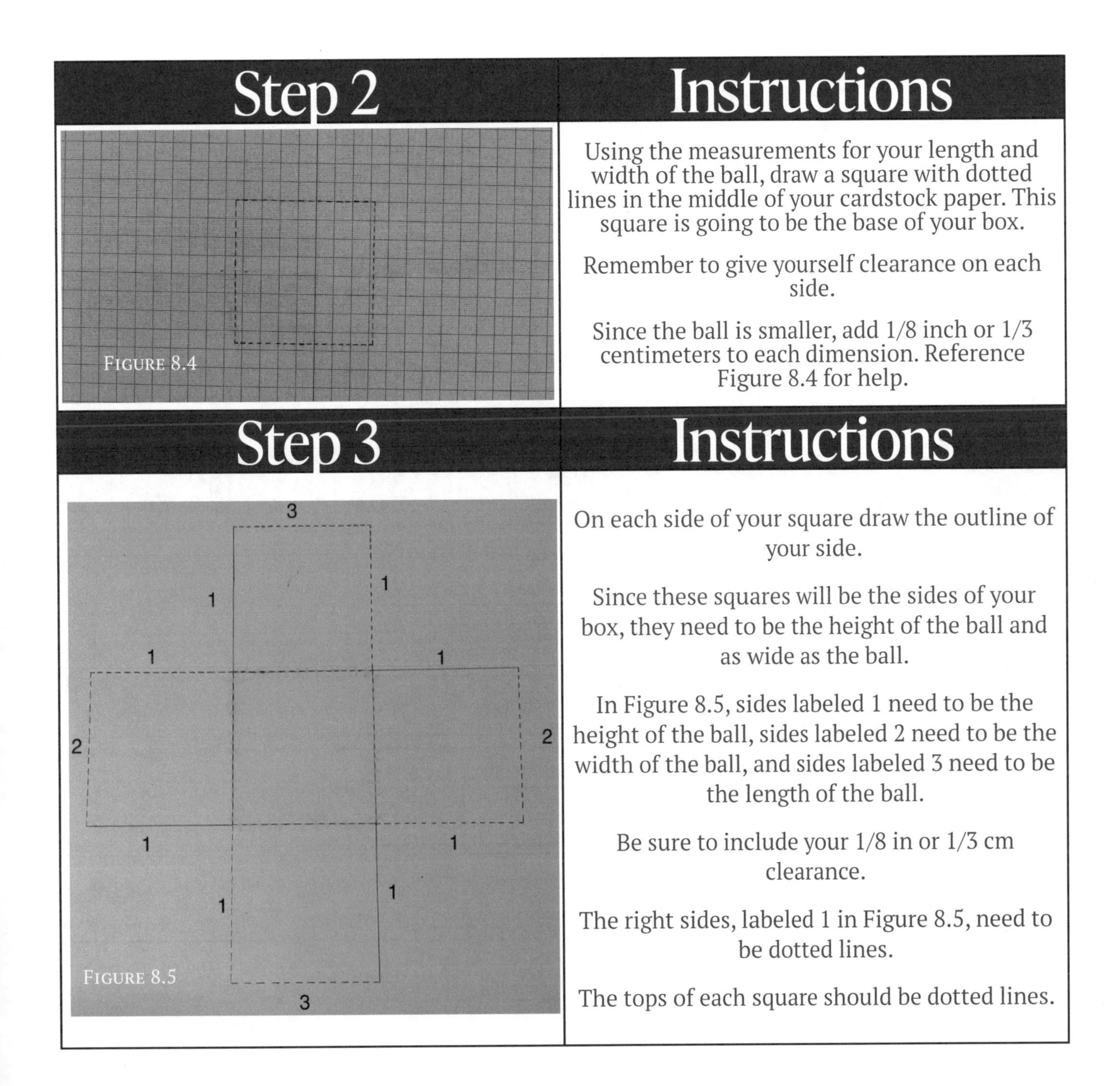

Step 2	Instructions
Figure 8.4	Using the measurements for your length and width of the ball, draw a square with dotted lines in the middle of your cardstock paper. This square is going to be the base of your box. Remember to give yourself clearance on each side. Since the ball is smaller, add 1/8 inch or 1/3 centimeters to each dimension. Reference Figure 8.4 for help.
Step 3	**Instructions**
3 1 1 1 1 2 2 1 1 1 1 3 Figure 8.5	On each side of your square draw the outline of your side. Since these squares will be the sides of your box, they need to be the height of the ball and as wide as the ball. In Figure 8.5, sides labeled 1 need to be the height of the ball, sides labeled 2 need to be the width of the ball, and sides labeled 3 need to be the length of the ball. Be sure to include your 1/8 in or 1/3 cm clearance. The right sides, labeled 1 in Figure 8.5, need to be dotted lines. The tops of each square should be dotted lines.

Package Design / Packaging

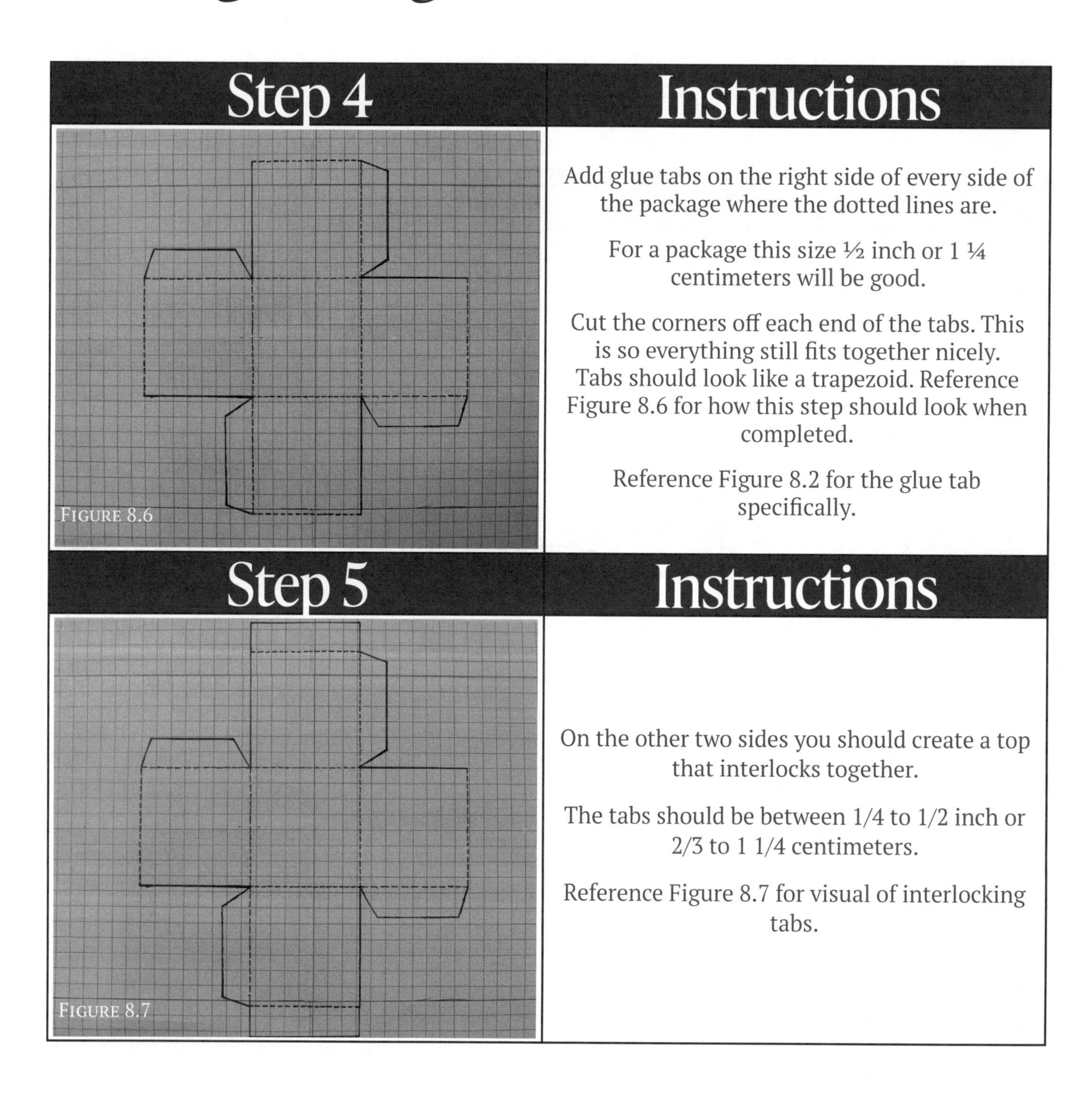

Figure 8.6

Figure 8.7

Package Design / Packaging

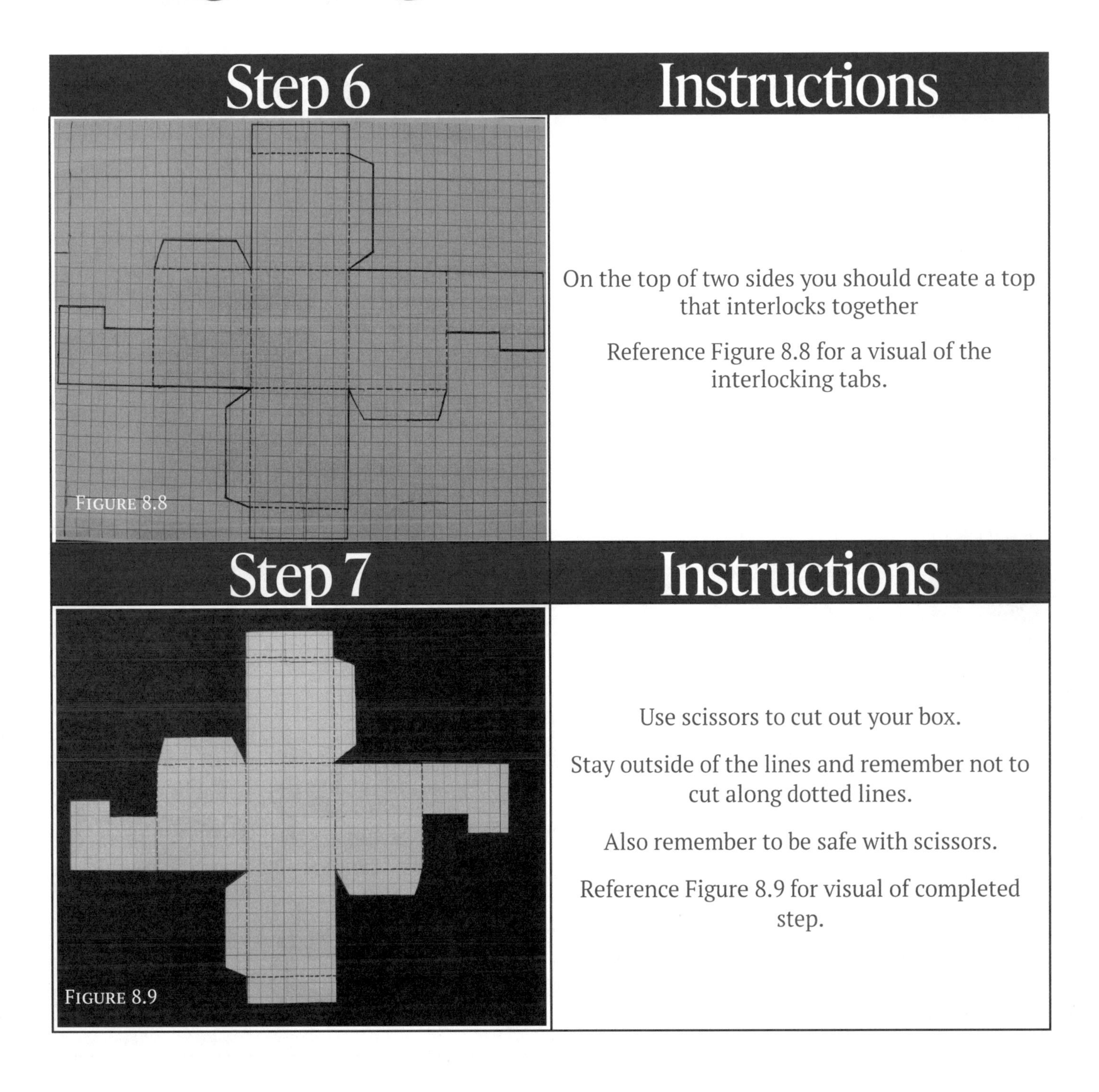

FIGURE 8.8

FIGURE 8.9

Package Design / Packaging

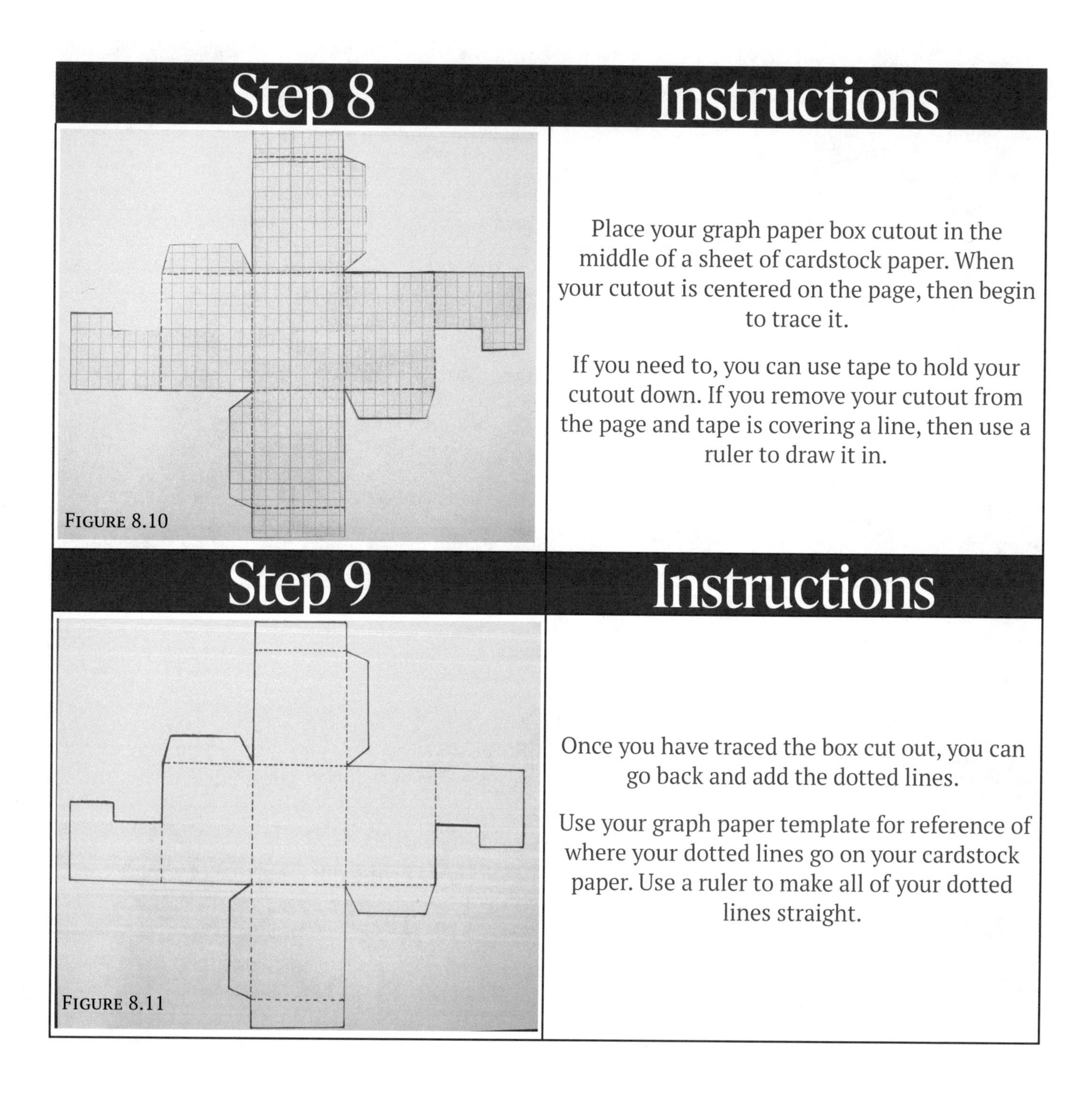

Step 8

FIGURE 8.10

Instructions

Place your graph paper box cutout in the middle of a sheet of cardstock paper. When your cutout is centered on the page, then begin to trace it.

If you need to, you can use tape to hold your cutout down. If you remove your cutout from the page and tape is covering a line, then use a ruler to draw it in.

Step 9

FIGURE 8.11

Instructions

Once you have traced the box cut out, you can go back and add the dotted lines.

Use your graph paper template for reference of where your dotted lines go on your cardstock paper. Use a ruler to make all of your dotted lines straight.

Package Design / Packaging

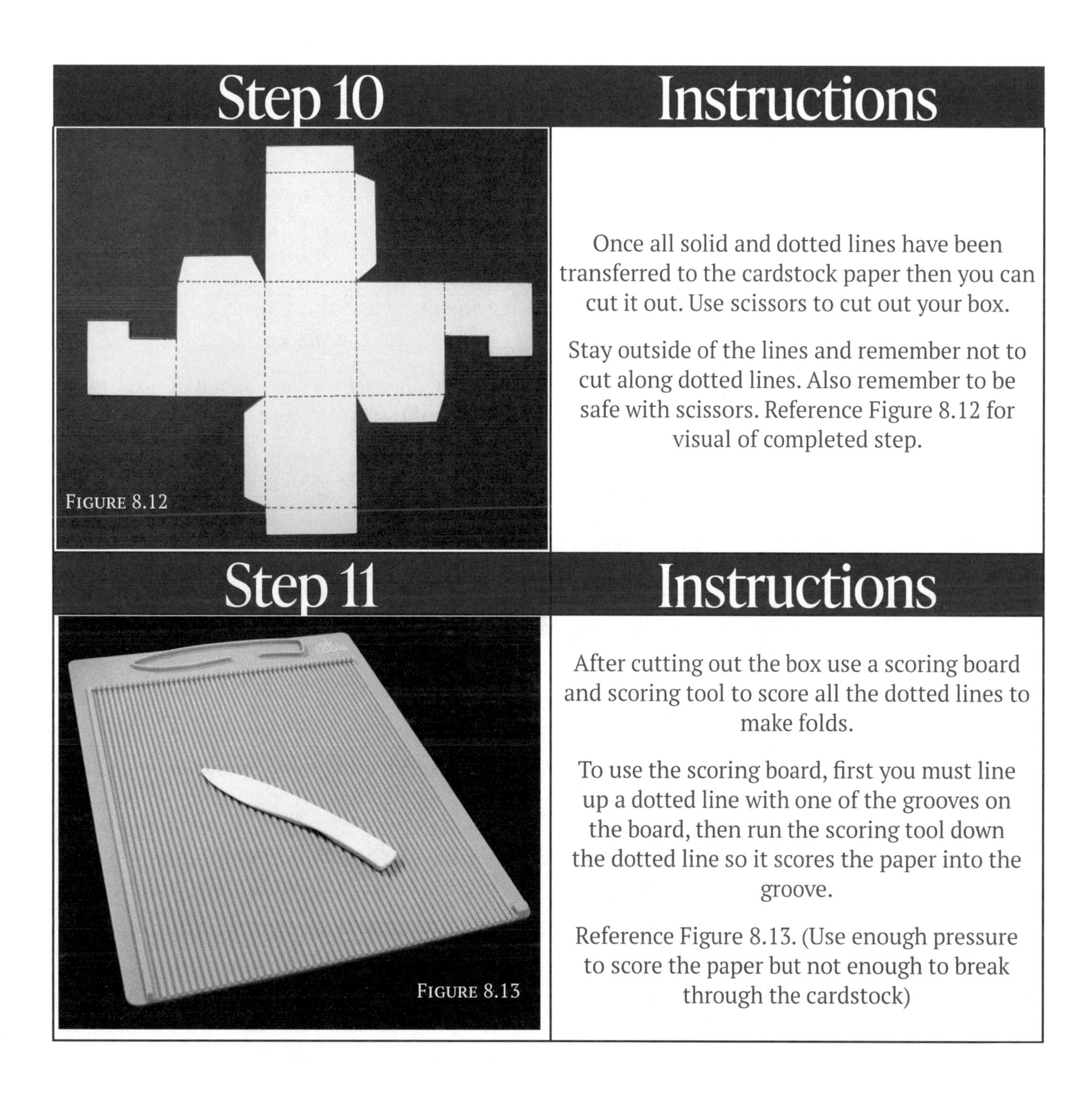

Step 10

FIGURE 8.12

Instructions

Once all solid and dotted lines have been transferred to the cardstock paper then you can cut it out. Use scissors to cut out your box.

Stay outside of the lines and remember not to cut along dotted lines. Also remember to be safe with scissors. Reference Figure 8.12 for visual of completed step.

Step 11

FIGURE 8.13

Instructions

After cutting out the box use a scoring board and scoring tool to score all the dotted lines to make folds.

To use the scoring board, first you must line up a dotted line with one of the grooves on the board, then run the scoring tool down the dotted line so it scores the paper into the groove.

Reference Figure 8.13. (Use enough pressure to score the paper but not enough to break through the cardstock)

Package Design / Packaging

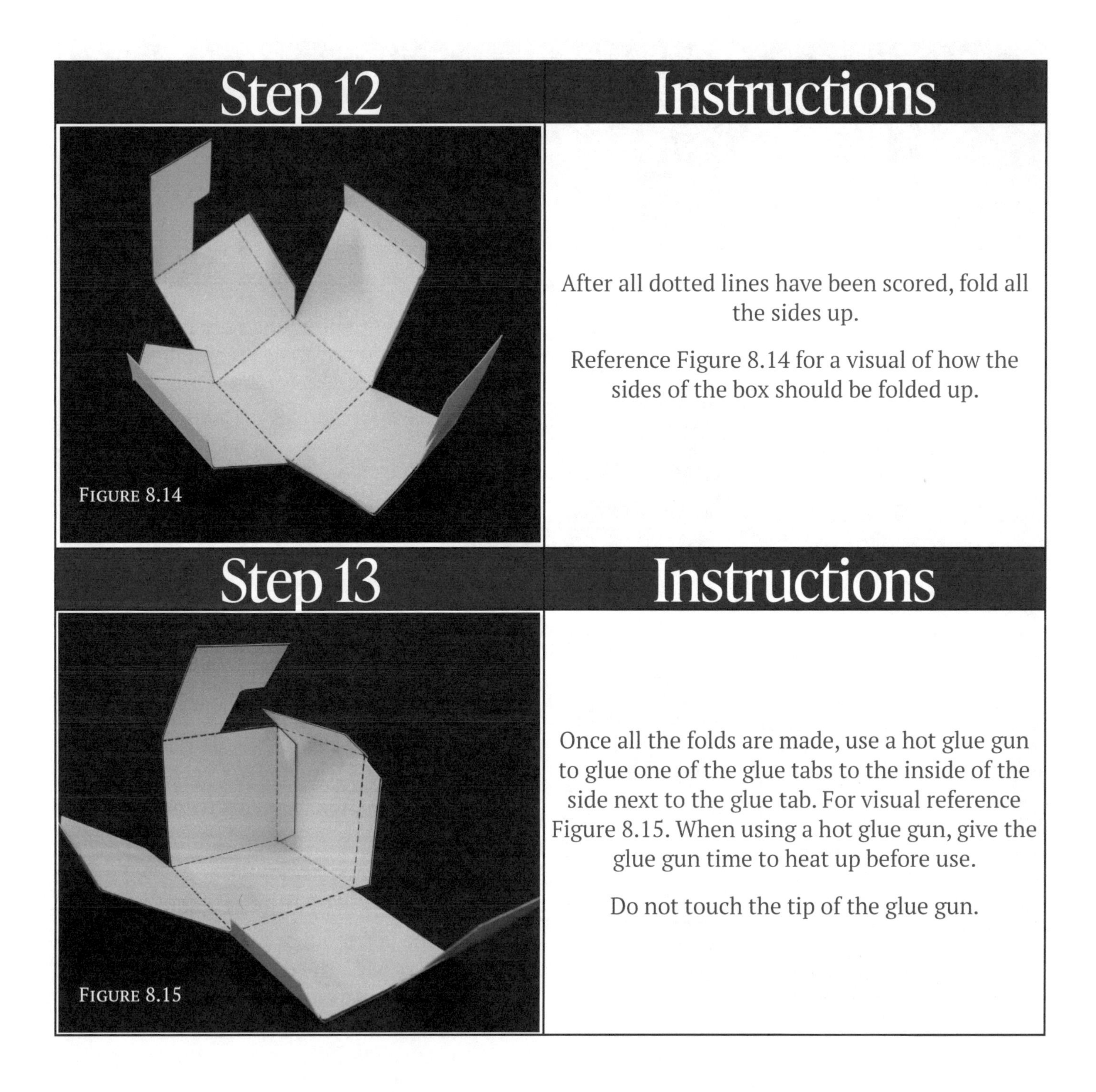

Step 12

FIGURE 8.14

Instructions

After all dotted lines have been scored, fold all the sides up.

Reference Figure 8.14 for a visual of how the sides of the box should be folded up.

Step 13

FIGURE 8.15

Instructions

Once all the folds are made, use a hot glue gun to glue one of the glue tabs to the inside of the side next to the glue tab. For visual reference Figure 8.15. When using a hot glue gun, give the glue gun time to heat up before use.

Do not touch the tip of the glue gun.

Package Design / **Packaging**

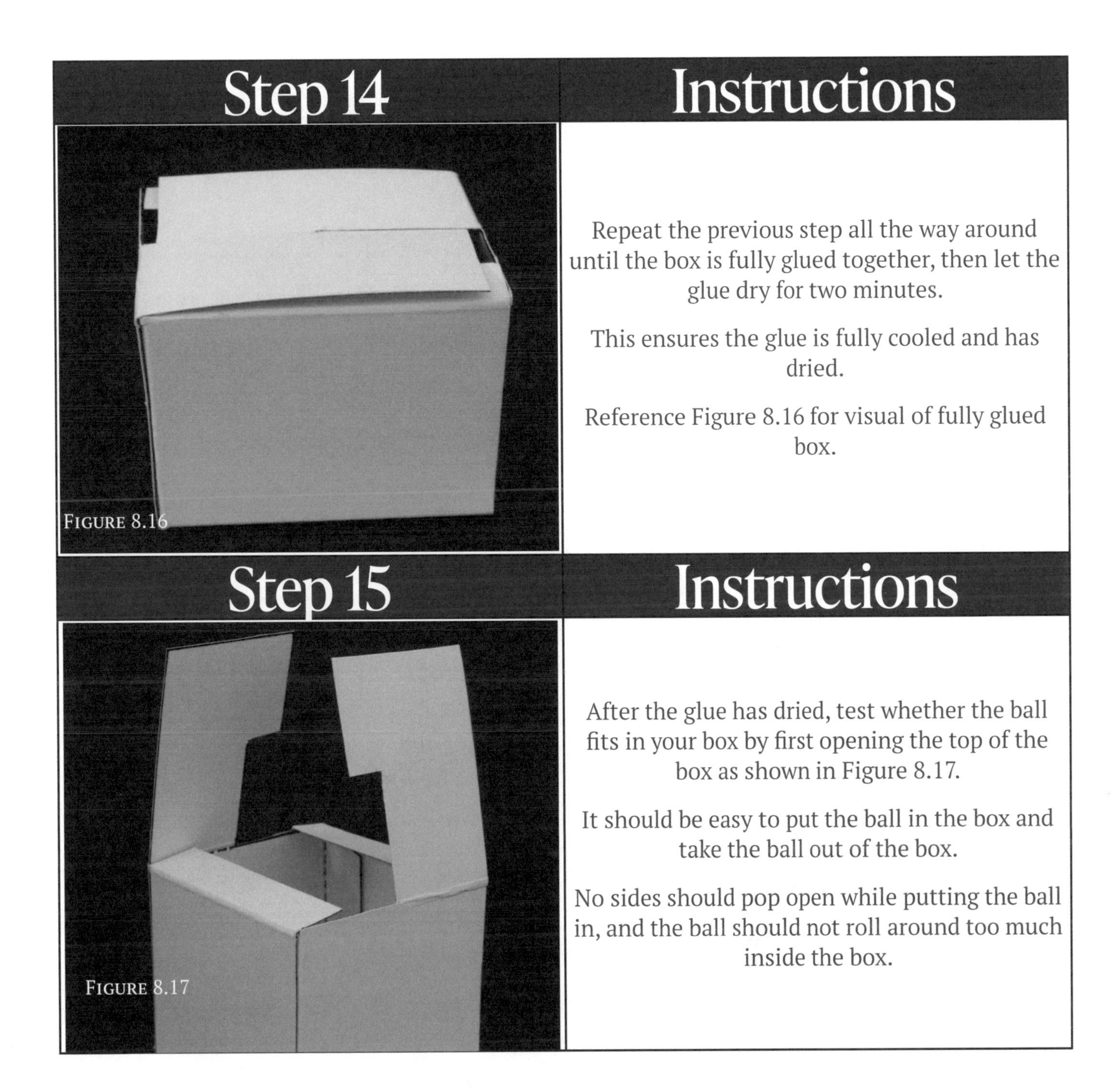

Step 14

FIGURE 8.16

Instructions

Repeat the previous step all the way around until the box is fully glued together, then let the glue dry for two minutes.

This ensures the glue is fully cooled and has dried.

Reference Figure 8.16 for visual of fully glued box.

Step 15

FIGURE 8.17

Instructions

After the glue has dried, test whether the ball fits in your box by first opening the top of the box as shown in Figure 8.17.

It should be easy to put the ball in the box and take the ball out of the box.

No sides should pop open while putting the ball in, and the ball should not roll around too much inside the box.

Package Design / Packaging

Troubleshooting	
Possible Problem #1 Problem: The ball does not fit. Possible Solution: Go back and compare your measurements from step one to other students, and create a new box if your measurements were off.	Possible Problem #2 Problem: The box will not hold together. Possible Solution: Do not be afraid to apply more glue on your box. Also hold the sides together until the glue completely dries before moving to the next side.

Observations & Conclusions / Packaging

Use the space below to draw an isometric sketch of your finished box with the ball in it. Analyze what went well with your design and build. Consider what could have gone better. Record your observations and conclusions below your sketch.

HYDRAULIC EXCAVATORS & BULLDOZERS

Photo by Dominik Vanyi

Hydraulic Robot: Properties of Materials

ROADMAP

- Have a clear understanding of the hydro robot you are going to be building and how it works.
- Understand how the strength of an object affects the construction of your project.
- Understand what flexibility is, how it makes structures unstable, and how it can be avoided.
- Know what force, load, and fulcrum mean and how they all work together in a lever.
- Know how weight distribution works and how it will affect the construction of your hydro robot.

Hydraulic Robot: Properties of Materials

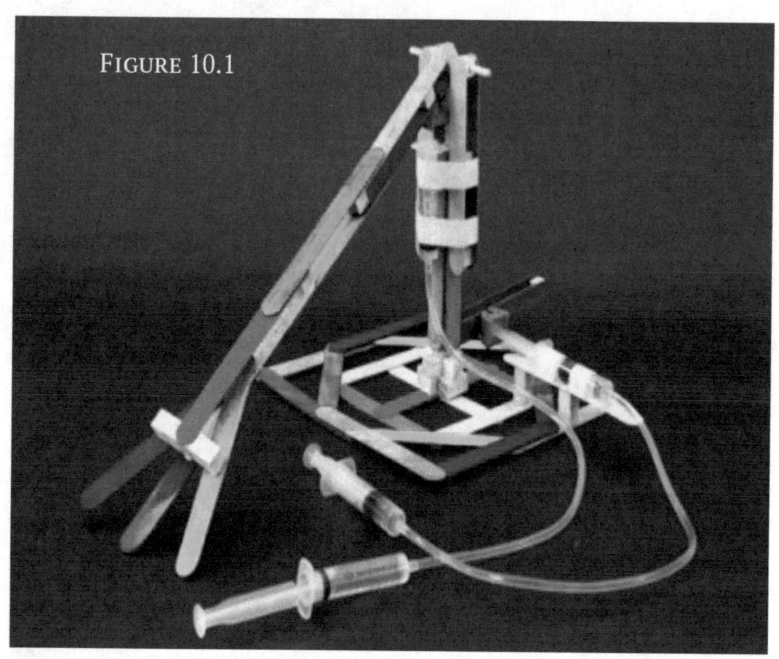
FIGURE 10.1

Introduction to the Hydraulic Robot Project

Over the course of the next few lessons, we are going to be covering properties of materials and forces by building a hydraulic robot out of craft sticks, wooden cubes, and plastic syringes. Your robot should have a stable base with a long arm that is able to reach and compete with other robots using a hydraulic syringe system.

Using this project as a guide, we are going to learn about different properties of materials as well as different kinds of forces. Once you are finished building your hydraulic robots, we will test the hydraulics and the weight distribution by engaging in a competition with other students.

The goal of the competition is to either flip the other robot upside down, or knock it out of its designated square. You will have an entire week to work on this project, but do not waste your time. There is much work to do!

The project's objective is to design a robot that fulfills the requirements of the following problem statement, criteria, and constraints.

Vocabulary

Problem Statement
A short, clear explanation of a problem you want to solve

Criteria
A list of the requirements a solution must meet to be considered successful

Constraints
A list of things that limit or control what you can do

Here is a quick refresher on the definitions of problem statement, criteria, and constraints:

Problem statement: A short, clear explanation of a problem you want to solve.

Criteria: A list of the requirements a solution must meet to be considered successful.

Constraints: A list of things that limit or control what you can do.

Problem statement: Build a hydraulic robot out of craft sticks and wooden cubes that uses water pumped through syringes to power the movement of the arm.

Let us look at the constraints and the criteria for building this project:

Criteria	Constraints
• Your robot must have fully functioning hydraulics. • Your robot must have a stable base so that the robot will not tip over by itself. • Your robot must be able to flip over another robot using its hydraulic powered arm.	Maximum material restrictions are: • 60 simple craft sticks or 50 colorful craft sticks • 30 wooden cubes without holes • 15 wooden cubes with holes • 1 dowel • 2 zip ties Additional constraints: • Your robot must contain at least 2 hydraulic systems • You may not cut any of your craft sticks • Your robot's base must be no bigger than 10 x 10 in

What is it and why is it important?

Today, we are going to cover the different physical properties of the materials you will be using to construct your hydro robot. Each material we will be using has many different properties, which are studied by material scientists and engineers. Two of these properties will be most important for you to understand are strength and flexibility. Additionally, physics will play a large role in the construction of your hydro robot. In today's lesson we will be analyzing the effect of weight distribution on your robot and its performance.

Strength

The **strength** of an object is its ability to withstand a great force or pressure without failing or breaking. In order to make our hydro robots as strong as possible with the materials provided, we cannot use only craft sticks because they are too flimsy and do not give enough strength to the structure of the robot. Instead, we must use a combination of craft sticks, wooden cubes, dowels and glue. This combination of materials will increase the overall strength of our robot. Before you begin any project, make sure the materials you are using will be strong enough to support its structure.

A·Z

Vocabulary

Strength
The ability of an object or substance to withstand great force or pressure

Flexibility
An object's ability to bend into unnatural positions without breaking.

Rigid
When an object is inflexible, unable to be bent, or unable to be forced out of shape

Force
A push or pull on an object that changes its state of rest or motion

Load
The heavy weight that is being lifted up due to the force exerted on the opposite side of the lever

Fulcrum
The point of the lever on which the beam pivots in a lever

Flexibility

Flexibility, in simple terms, is an object's ability to bend without breaking. Most building scenarios call for the product to be **rigid** and not flexible. The craft sticks you will be using for this project are mostly rigid, but are still flexible enough to make them prone to bending over any external weight placed on them.

FIGURE 10.2
STRENGTHENED CRAFT STICKS USING A CRISS-CROSS PATTERN

However there are several methods you can use to minimize the flexibility of your project to make it more sturdy. One way to strengthen the structure of your robot and reduce flexibility is to make craft sticks run across each other in a criss-cross shape, which is shown in Figure 10.2.

Another way to reduce the flexiblity of craft sticks is to stack them on layer them on top of each other. Always make sure that your building materials are not too flexible before you begin construction so that they do not weaken the overall structure of your project.

Quick Physics Lesson

Before we begin talking about weight distribution, it is important to understand a few simple physics concepts first. Have you ever looked at a seesaw before and wondered how it works? Well, it is actually pretty simple. Whenever you push down on one side of the seesaw, you are exerting a force on that side. A **force** is a push or pull on an object that changes its state of rest or motion. The load is typically on the other side of the seesaw, as you can see in Figure 10.3. The **load** is the heavy weight that is being lifted up due to the force exerted on the opposite side. Finally, the **fulcrum** is the point on which the beam pivots. Changing the location of the fulcrum is an easy way to change how much force is required to move the load. This setup shown in Figure 10.3 is very important in physics and is more commonly known as a simple lever.

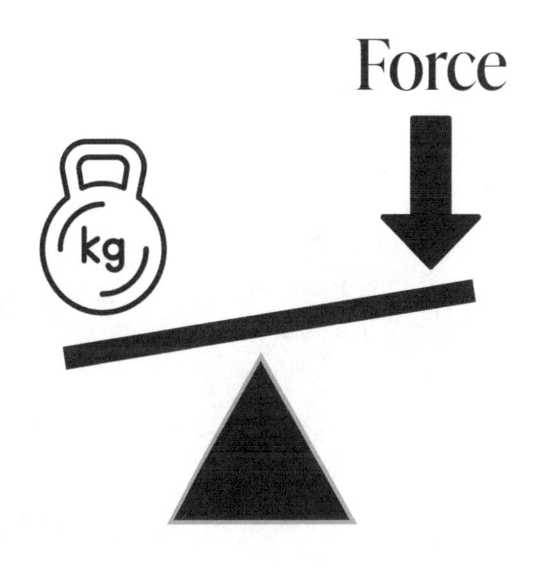

FIGURE 10.3
A SIMPLE LEVER

Weight Distribution

One of the most important things to take into consideration when building something is the **weight distribution** of the project you are building. Recall from Lesson 7 the concepts of mass and weight. Mass is the measurement of how much matter is in an object. Weight, on the other hand, is the measurement of how much gravitational force is exerted on an object and is the reason objects fall to Earth and other large gravitational bodies.

Weight distribution, then, is how much weight is imposed on the ground at different points on the object. Let's use the hydraulic robot as an example. If the arm of your robot is extremely long and reaches past the front edge of your base, the weight will be distributed towards the front of your base. If you rotate the arm to point towards the left side of your base, then the weight will be distributed towards the left side of your base.

When your robot is standing upright, the point at which the total weight of the object is imagined to be concentrated is the center of gravity. It is also the point around which the entire object would be able to balance perfectly. Once you find the **center of gravity** on your hydro robot, you can use that information to strategize how you will use it in your fight with another robot. For example, if the center of gravity of your robot is at the front of your robot, you should be careful not to lunge forward at the other robot too much because your robot will be most likely to fall over forwards.

Another way we can look at weight distribution is shown in Figure 10.4. This diagram shows a crane which has a similar structure to the robot you will be building. In your project the robot's arm will be a claw used to push other robots and will be capable of moving up and down, unlike the fixed crane arm. As you can see in the diagram, the arm is attached to the central beam by a pivot. This means that the arm is able to rotate or spin in different directions.

Vocabulary

Weight Distribution
The amount of weight imposed on the ground at different points by an object

Weight
The amount of gravitational force exerted on an object expressed in ounces or grams

Center of Gravity
An imaginary point in a body of matter where, for convenience in certain calculations, the total weight of the body may be thought to be concentrated

Counterweight
The ability of an object or substance to withstand great force or pressure

Pivot
The point around which the arm rotates in a simple lever.

You may be surprised that massive cranes like these do not fall over due to the size of the arm. When the weight of the arm is evenly distributed they do not fall over. The load, as shown in the diagram, is the object which the crane is carrying and is usually very heavy. A **counterweight** is attached to the back of the arm in order to balance out the weight distribution.

Without the counterweight, the load hanging from the arm would create an uneven weight distribution, which would throw the crane off balance and possibly tip over the entire structure. Finally, the **pivot** is the point around which the arm rotates.

In the scenario of the crane, the counterweight is the force, because it is lifting up the load, the load remains the same, and the pivot is the fulcrum because it is the point on which the arm pivots. Figure 10.4 will help you understand this.

Now that you understand these three properties of materials (strength, flexibility, and weight distribution), let us move on to the activity.

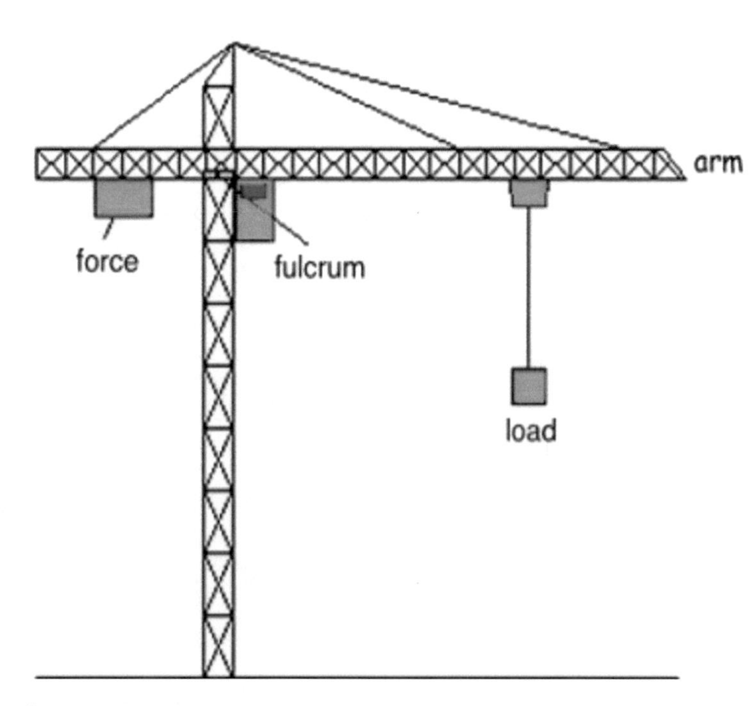

FIGURE 10.4
SIMPLE LEVER

ACTIVITY

Crossword Puzzle / Properties of Materials

Crossword Puzzle / Properties of Materials

Across	Down
2. A weight that is used to balance out a load in order to make lifting the load more efficient	1. A list of things that limit or control what you can do
5. The heavy weight that is being lifted up due to the force exerted on the opposite side of the lever	3. A list of the requirements a solution must meet to be considered successful
8. A short, clear explanation of a problem you want to solve	4. An object's ability to withstand a great force or pressure
9. The amount of weight imposed on the ground at different points by an object	6. A push or pull on an object that changes its state of rest or motion
10. The point of the lever on which the beam pivots	7. An object's ability to bend into unnatural positions without breaking
	11. When an object is inflexible, unable to be bent, or unable to be forced out of shape

A perfect example of structure that uses the physical properties of materials that we covered today is the Alcantara Bridge (Figure 10.6). This bridge was built across the Tagus River under the Roman emperor, Trajan, around 130 or 140 BC. It is almost 2000 years old and is still in use today. As you can see, it would be very difficult to build something like this without considering strength and flexibility of the materials as well as the weight distribution of the entire structure. Obviously, this bridge was built with massive stone blocks which are both strong and rigid, so the building materials were not something to worry about. However, how do you construct such a great structure without it collapsing? This enormous bridge is still standing today because it uses multiple arches between evenly spaced pillars to spread the weight across the entire bridge so that it can support something as heavy as a semi-truck.

HYDRAULIC EXCAVATOR

Photo by Fly&I

LESSON

Hydraulic Robot: Forces

ROADMAP

- Understand gravity and gravitational force.
- Understand the concepts of resistance and friction.
- Learn the differences between static, sliding and rolling friction.
- Recognize hydraulic systems around you–.
- Identify parts of a hydraulic jack.

Hydraulic Robot: Forces

What is it and why is it important?

In the previous lesson, we briefly talked about the concept of force and the role it plays in the weight distribution of an object. A force is any sort of push or pull on an object and there are many different kinds. Today we are going to cover the three most important ones. Gravity, friction, and hydraulics are all important to understand because they will affect the performance of your robot.

Gravity

Gravity is a force of attraction between objects, but for this lesson what you need to know is that gravity pulls things to the ground. For example gravity is the force that causes a pencil to fall to the floor when you drop it and is the reason why planets stay in their orbits around the Sun.

Although at first you may think that gravity is insignificant and will have little effect on the performance of your hydro robot, it actually plays a much bigger role than you might think. Gravity is what causes the robot to fall over if it is unstable or gets hit by another robot.

As we discussed in the previous lesson, weight is the measurement of how much **gravitational force** is exerted on an object. That means that larger, heavier objects will be more affected by gravity than smaller, lighter objects. The arm of your robot will consist of craft sticks and wooden cubes which are small and light on their own. However, when you attach more and more pieces to the arm of your robot, the weight will add up and become heavier and heavier until it falls over due to gravity.

Figure 11.1 shows an accident that happened because the mobile crane was carrying too much weight on its arm. Similarly, if you add too many craft sticks and cubes to the arm of your robot, your robot will either not be able to lift the arm, or it will cause your entire hydraulic robot to fall over, just like the crane in the picture.

FIGURE 11.1
GRAVITATIONAL FORCE AT WORK

Vocabulary

Gravity
The force that pulls objects toward the earth

Gravitational Force
The force exerted on an object due to gravity

Friction
The resistance that occurs when two surfaces come into contact with each other

Resistance
The force that slows or stops the progress of an object in motion.

Static Friction
The frictional force that keeps an object in a static (still) position

Sliding Friction
The frictional force that occurs while one object slides along another

Rolling Friction
The resistance that occurs when a wheel or ball rolls across a surface

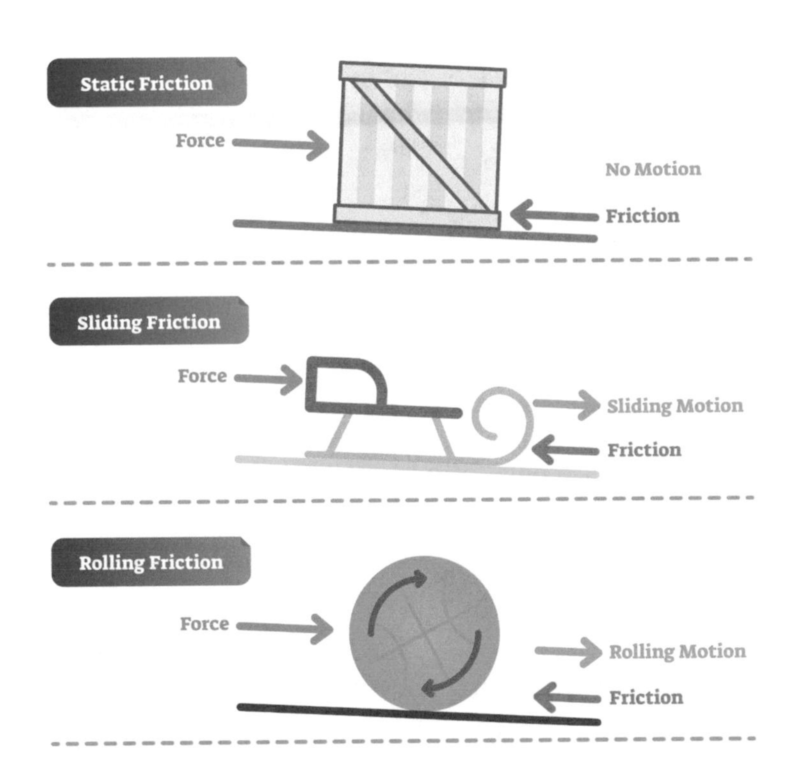

FIGURE 11.2
DIFFERENT TYPES OF FRICTION

Friction

The force of **friction** plays a major role in the performance of your hydro robot. In order to understand friction, you must know what resistance is. **Resistance** is the force that slows or stops the progress of an object in motion. Friction, in simple terms, is the resistance that occurs when two surfaces come into contact with each other. There are many different kinds of friction, but today we will be covering the three most common types shown in Figure 11.2: static friction, sliding friction, and rolling friction. A brief description of each is given below.

Static friction: The fricitional force that keeps an object in a static (still) position.

Sliding friction: The frictional force that occurs while one object slides along another.

Rolling friction: The resistance that occurs when a wheel or ball rolls across a surface.

Today, we will be focusing on sliding friction because it is most relevant to the performance of the hydro robot.

As mentioned above, sliding friction is the resistance created when two objects slide against each other. The resistance created by sliding friction is what makes some objects harder to slide than others. Have you heard about cars losing control and spinning out on wet roads? This happens more often on wet roads because water reduces friction. On a dry road, there is more resistance between the car tires and the road, therefore, there is more friction.

The same concept can be applied to the base of the hydro robot. Craft sticks have a very low resistance to other objects, so they slide along the ground very easily. However, if we were to add drops of glue to the bottom of the base and wait for them to dry, we would be left with a soft rubbery padding on the bottom of the craft sticks. The dried glue has much more resistance to other surfaces than craft sticks do, which means there is more friction between the surface of the table and the dried glue. This will make it much more difficult to slide your base across the table so you will have a better chance of staying within your designated square while competing against other robots. This use of friction could be the difference between victory and defeat in your competition.

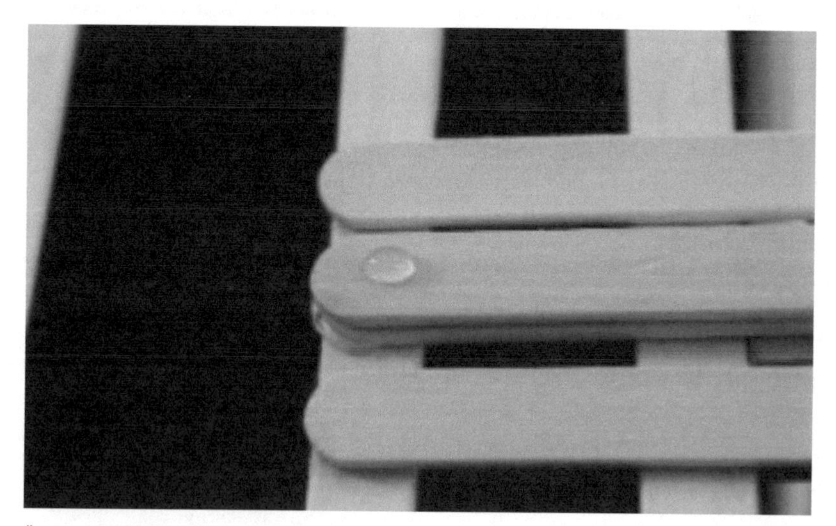

FIGURE 11.3
GLUE DOTS INCREASE SLIDING FRICTION

Vocabulary

Hydraulic System
A system that pumps a fluid through a series of components to convert liquid pressure into mechanical energy

Hydraulics
Another term used to describe hydraulic systems

Liquid Pressure
The force per unit area exerted by a liquid in all directions

Hydraulic Systems

A **hydraulic system** is a system that uses liquid to move or lift loads. Usually, water or oil is pumped through a tube to convert liquid pressure into mechanical energy. The most important thing to understand when learning about **hydraulics** is liquid pressure. **Liquid pressure**, simply put, is how much force per unit area is exerted by a fluid in all directions. This force can be caused by anything that compresses, or squeezes, the liquid. When dealing with hydraulics, the liquid is usually compressed in a tube, and has no room to expand, so it pushes on the ends of the tube creating a very strong force. This force can be used to move very large objects such as elevators. Although these definitions of hydraulics and liquid pressure may seem confusing at first, you will be able to see how it all works once you get started on your robot.

Some of the most common machines that use hydraulics are car shock absorbers, construction equipment, roller coasters, and adjustable chairs.

FIGURE 11.4
HYDRAULICS IN A CAR

FIGURE 11.5
HYDRAULICS IN HEAVY MACHINERY

FIGURE 11.6
HYDRAULICS IN A ROLLERCOASTER

FIGURE 11.7
HYDRAULICS IN AN ADJUSTABLE CHAIR

Vocabulary

Valve
A mechanical device used to regulate the flow of a liquid or gas.

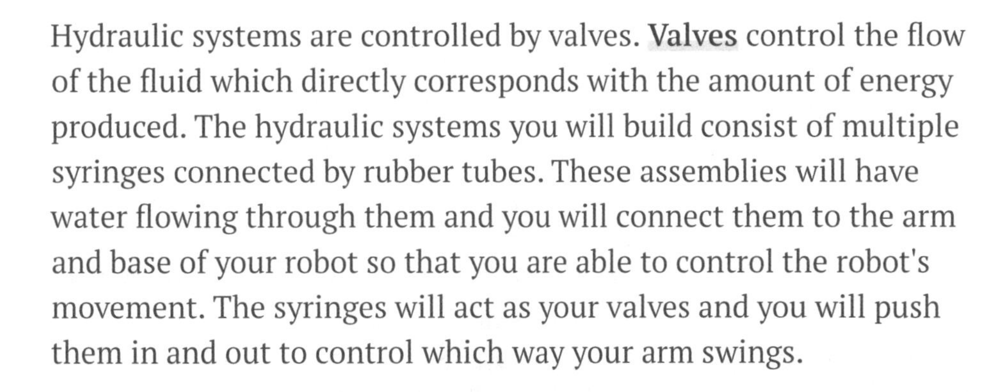

Hydraulic systems are controlled by valves. **Valves** control the flow of the fluid which directly corresponds with the amount of energy produced. The hydraulic systems you will build consist of multiple syringes connected by rubber tubes. These assemblies will have water flowing through them and you will connect them to the arm and base of your robot so that you are able to control the robot's movement. The syringes will act as your valves and you will push them in and out to control which way your arm swings.

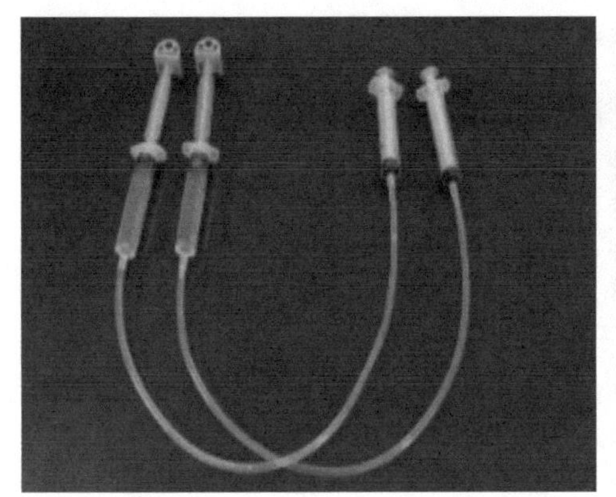

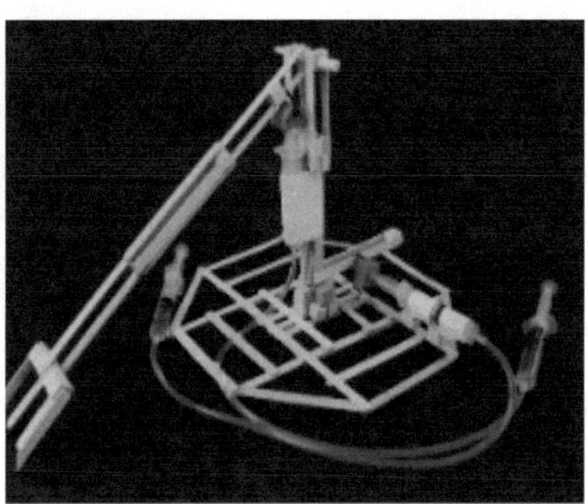

FIGURE 11.8 AND 11.9
HYDRAULIC ASSEMBLY BEFORE AND AFTER ATTACHING TO A ROBOT

Now that you understand these three forces that will affect the construction of your hydro robot (gravity, friction, and hydraulics), let's move on to the activity.

FIGURE 11.10
VALVES IN INDUSTRY

Lesson Review / Forces

Instructions: Use the vocabulary words found throughout this lesson to fill in the blanks, then use your answers to complete the word search on the next page.

1. **The attraction between all objects that have mass; the force that pulls objects toward the earth: ________________________**
2. **The force exerted on an object due to gravity: ________________________**
3. **The force that slows or stops the progress of an object in motion:**
   ____
4. **The resistance that occurs when two surfaces come into contact with each other:**
   ____
5. **The amount of force exerted on a certain amount of fluid: ________________________**
6. **A system that pumps a fluid through a series of components to convert liquid pressure into mechanical energy: ________________________**
7. **Another way of describing a hydraulic system: ________________________**
8. **A mechanical device used to control the flow of a fluid in a hydraulic system: ________**

The first hydraulics were used by a Swiss mathematician named Daniel Bernoulli in 1738. Bernoulli studied liquid behavior in enclosed tubes, more specifically, the speed at which blood flows through blood vessels.

Using his knowledge of liquid pressure as well as the speed at which liquids flow, Bernoulli theorized that these forces could be harnessed and used to generate power. Bernoulli experimented with these ideas in water mills and pumps. Today, Bernoulli is known worldwide as the first man to experiment with the use of hydraulics.

Word Search / Forces

Instructions: Use the answers to the exercise on the previous page to complete the word search below.

M	W	Z	R	M	E	S	R	J	T	W	F	K	G	Y	V	X	G	A	S
A	D	W	E	R	C	M	L	S	L	L	R	J	O	O	F	N	X	U	U
P	V	V	S	M	B	U	P	V	A	L	V	E	C	Y	F	S	W	C	Y
Y	T	P	I	E	C	G	T	L	J	R	N	I	I	S	E	J	X	I	M
S	U	V	S	A	Q	G	C	G	R	O	X	T	M	N	C	I	Y	B	E
Z	I	H	T	D	L	I	Q	U	I	D	P	R	E	S	S	U	R	E	T
V	Q	K	A	C	H	M	Y	T	A	X	O	Z	U	Y	U	U	O	E	S
S	A	S	N	O	D	U	C	G	D	Y	I	V	E	E	P	R	X	W	Y
C	X	R	C	C	A	I	T	K	T	Q	J	P	M	Y	J	Y	G	H	S
K	W	W	E	O	R	M	X	I	Z	L	V	M	Y	X	T	Y	P	A	C
C	B	B	O	F	R	A	V	B	C	Z	Y	M	Z	M	R	F	A	N	I
M	P	D	W	C	W	A	U	U	D	O	F	O	A	O	C	Y	G	V	L
F	Z	I	K	X	R	Z	Y	H	Y	D	R	A	U	L	I	C	S	X	U
N	Z	U	V	G	J	F	Z	C	F	Z	Q	T	O	I	R	H	P	J	A
L	A	Q	S	I	I	Q	Q	R	Q	R	V	M	J	T	U	A	N	B	R
O	G	H	G	E	K	C	C	N	L	K	G	H	Z	M	R	C	G	R	D
G	R	A	V	I	T	A	T	I	O	N	A	L	F	O	R	C	E	Q	Y
K	M	O	B	V	H	J	W	Y	V	E	W	S	W	I	W	Y	W	M	H

Component Identification / Forces

Label the diagram of this Hydraulic Jack. Use the list of parts below for clues and fill in the blanks.

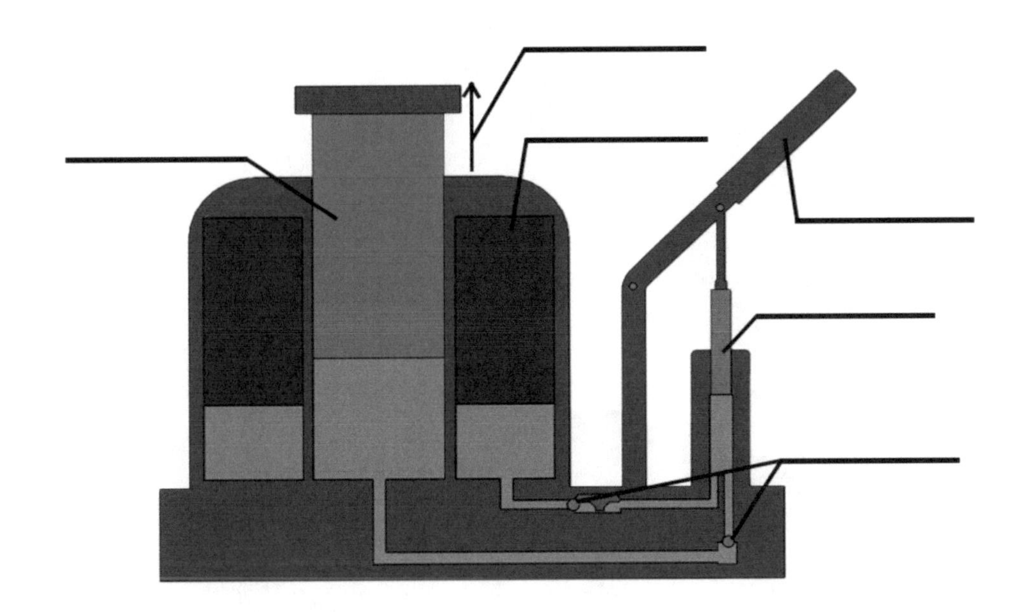

How does a Hydraulic Jack work?

Pump Handle: When you move the handle of the jack up and down, it pushes a small piston into a cylinder filled with hydraulic fluid (usually oil).

Small Cylinder: As the handle is pumped, the small piston pushes the hyrdraulic fluid through a one-way valve into a larger cylinder.

Larger Cylinder: The fluid moves from the small cylinder to the larger cylinder, which has a bigger piston inside it.

Lifting: Because of the difference in size between the small and large pistons, the pressure created in the small cylinder (by pumping the handle) generates a much larger force in the large cylinder. This force is what lifts the heavy object.

Check Valves: The one-way valves ensure that the fluid only moves in one direction (from the small cylinder to the larger cylinder) which makes sure that the lifting action continues each time you pump the handle.

Reservoir Tank: This tank holds the excess hydraulic fluid that has not yet been moved with the small cylinder to lift the large cylinder.

HYDRAULIC EXCAVATORS

12

LESSON

Hydraulic Robot: Project Planning

ROADMAP

- Understand how planning the design of a project works and why it is important.
- Be able to sketch orthographic projections which include multiple views of the base, main structure, and arm of the robot.
- Be able to sketch an isometric projection of the entire structure of the robot.
- Learn to share ideas, collaborate and work together as a team.

Hydraulic Robot: Project Planning

What is it and why is it important?

Today we are going to use our engineering notebooks to plan out the overall design of the hydro robot. The process of planning out a project can be tedious and even boring for some. However, **project planning** is one of the most important steps when beginning any project. A project plan usually consists of an isometric projection of your project and as many orthographic projections as needed to fully represent your project. (See lessons 4 and 5 for isometric and orthographic drawing basics). These drawings will help the builder to understand exactly what it is you are building as well as the exact dimensions of each component. Your project plans should include one isometric projection of the entire robot, and multiple orthographic projections of the base, main structure, and arm of the robot.

How does it work?

Before you begin planning out your project, it is important to fill out all the important information on each page of your notebook. Once you get in the habit of doing this, it will make it easier for you to find each project when you have your own engineering notebook. It will also make your notebook look more professional. It is also important to note your problem statement, constraints, and criteria. These will help guide you when planning the designs for your hydro robot.

Remember that when you are sketching isometric and orthographic projections, you must always use a ruler. Also, make sure your measurements are precise and **proportionate**. You will begin drawing out the design of your base using an orthographic projection of the top view because it will remain flat on the ground.

If your robot looked like the one shown in Figure 12.1, your orthographic and isometric projections would be as follows.

An example of what a plan for the base might look like is shown in Figure 12.2.

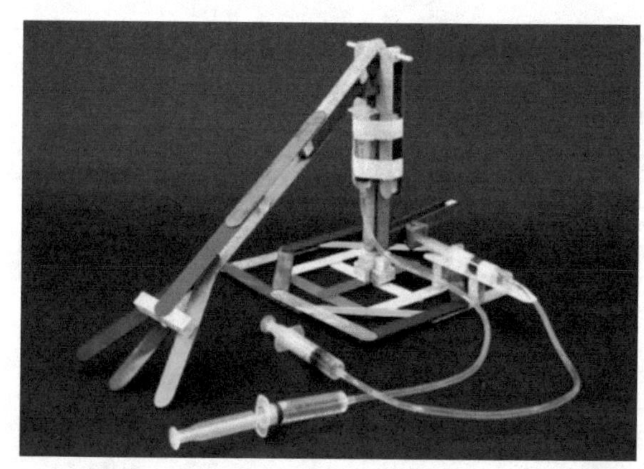

FIGURE 12.1
AN EXAMPLE OF A HYDRAULIC ROBOT

Vocabulary

Project Planning
Determining the actual steps to complete a project

Proportionate
When sizes are in balance so pieces fit together properly

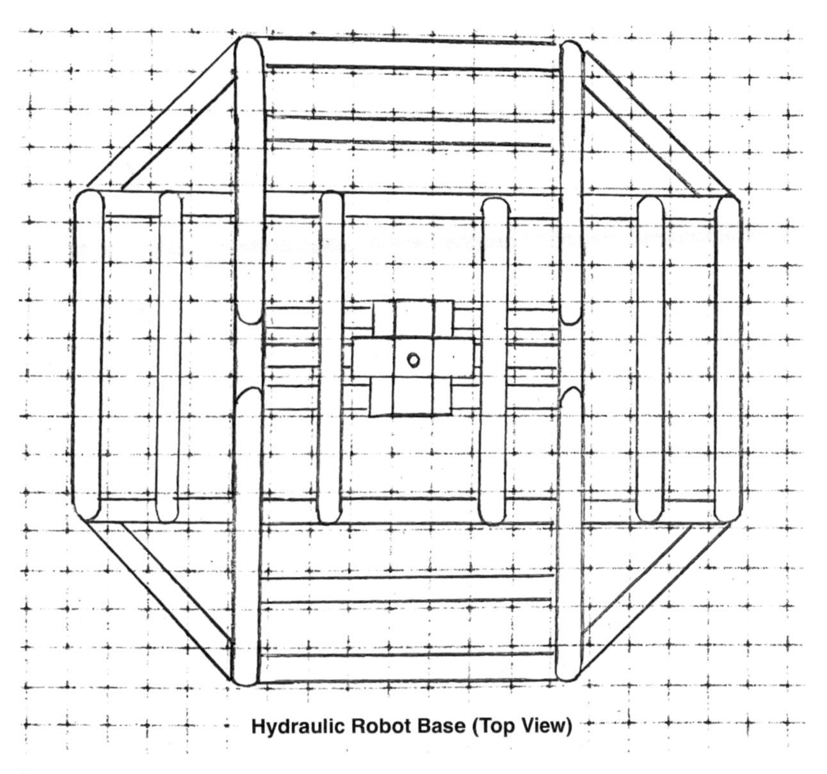

FIGURE 12.2
TOP DOWN VIEW OF A HYDRAULIC ROBOT

Next, you will plan out what the main structure of your robot will look like. The plan for the main structure of your robot will require multiple orthographic projections. An example of what a plan for the main structure might look like is shown in Figure 12.3.

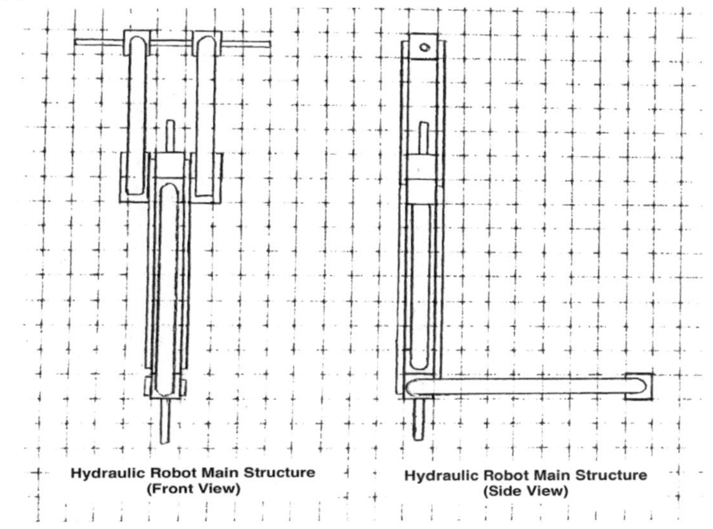

FIGURE 12.3
MAIN STRUCTURE FRONT AND SIDE VIEW

Next, you will plan out what the arm of your robot will look like. The plan for the arm of your robot will require multiple orthographic projections. An example of what a plan for the main structure might look like is shown in Figure 12.4.

Vocabulary

Team Work
Completing a task together to achieve a common goal

Collaboration
The action of working with someone to produce or create something

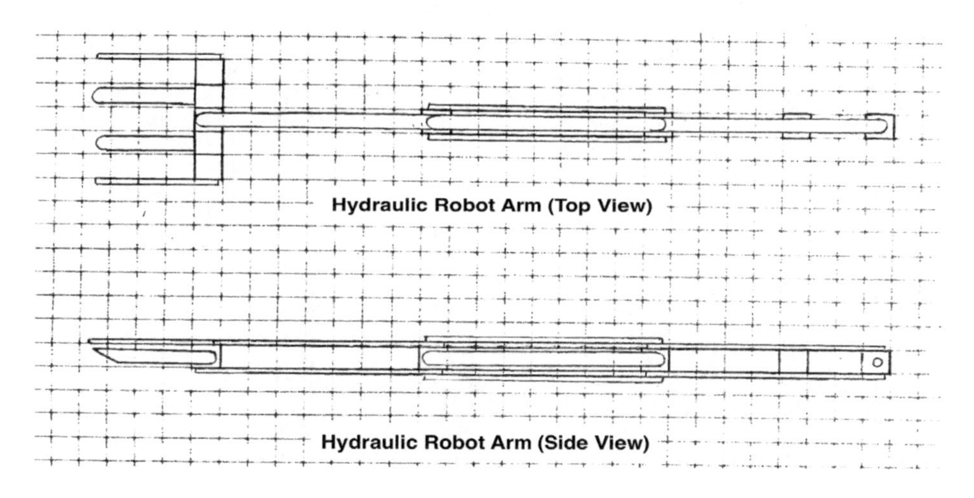

FIGURE 12.4
TOP AND SIDE VIEW OF HYDRAULIC ROBOT ARM

Next, you will plan out the entire structure of your hydro robot using an isometric projection. An example of what an isometric projection for this might look like is shown in Figure 12.5.

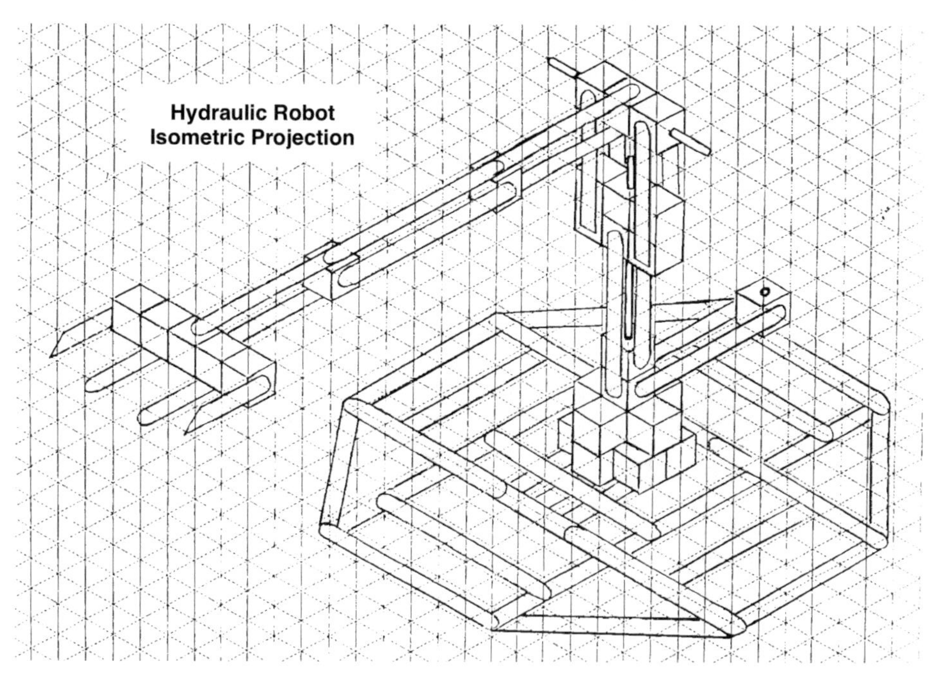

FIGURE 12.5
ISOMETRIC PROJECTION OF HYDRAULIC ROBOT

Finally, you will decide how you will attach the hydraulic system to your robot in a way that will make it run smoothly. Figures 12.6 - 12.8 show multiple views of one possible method for attaching your hydraulic system to your robot.

FIGURE 12.6
ATTACHING HYDRAULICS - VIEW 1

FIGURE 12.7
ATTACHING HYDRAULICS - VIEW 2

You will be planning, building, and testing your hydraulic robot in groups of two. Your teacher will assign you your partner. For the planning section of this project, make sure you discuss your robot design with your partner and agree on ONE design. This requires **team work** and **collaboration**. Then, you may work together drawing out project plans. For example, one of you could draw projections of the base of the robot while the other person draws projections of the robot arm. Continue using this system until you have completed all plans for your hydraulic robot.

FIGURE 12.8
ATTACHING HYDRAULICS - VIEW 3

Plan a Hydraulic Robot / **Project Planning**

For this experiment, you will need the following materials:

- Sharpened pencil & eraser
- Ruler with measurements
- Partner for hydraulic robot project

Problem Statement:

Build a hydraulic robot out of craft sticks and wooden cubes that uses water pumped through syringes to power the movement of the arm.

Criteria	Constraints
• Your robot must have fully functioning hydraulics • Your robot must have a stable base so that the robot will not tip over by itself • Your robot must be able to flip over another robot using its hydraulic powered arm	Maximum material restrictions are: • 60 simple craft sticks or 50 colorful craft sticks • 30 wooden cubes without holes • 15 wooden cubes with holes • 1 dowel • 2 zip ties Additional constraints: • Your robot must contain at least 2 hydraulic systems • You may not cut any of your craft sticks • Your robot's base must be no bigger than 10 in x 10 in

Plan a Hydraulic Robot / Project Planning

Step	Explanation
Step 1	Before you begin drawing out any sketches of build designs, take a few minutes to brainstorm with your partner. Take out your materials and look at them and measure them. Measure out a 10 in by 10 in area so you understand the size constraint for the base. How many craft sticks would you use to make a 10 in by 10 in base? Record your observations on the graph paper following the instructions.
Step 2	Discuss with your partner what kind of design you want to use for your robot, consider the amount of materials you have and the measurements you have taken and agree on ONE design.
Step 3	Plan out what the base of your hydraulic robot will look like using an orthographic projection of the top view. Make sure all of your plans will allow you to stay within the constraints of materials or you will run out of craft sticks and blocks.
Step 4	Plan out what the main structure of your hydraulic robot will look like using multiple orthographic projections.

Plan a Hydraulic Robot / Project Planning

Step	Explanation
Step 5	Plan out what the arm of your hydraulic robot will look like using multiple orthographic projections. Make sure to stay within the constraints of materials.
Step 6	Draw an isometric projection of your entire hydro robot.
Step 7	Decide how you want to attach the two hydraulic assemblies to your robot.

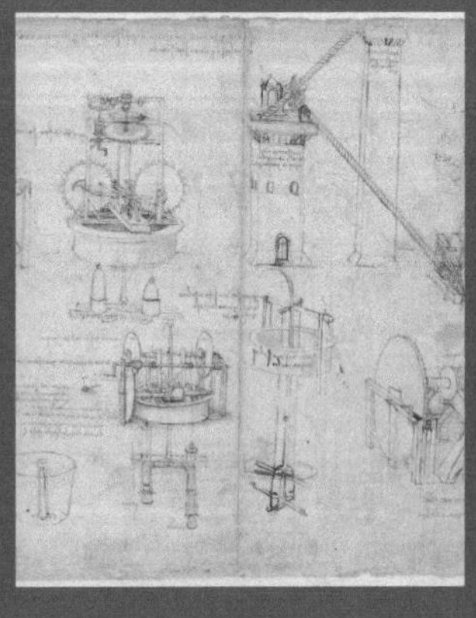

Did you know that the world-famous Renaissance artist and scientist, Leonardo da Vinci (painter of the Mona Lisa), used an engineering notebook for his project planning? It was nothing like the fancy new books you are using now, but a stack of rough paper bound together in a leather covering. The notebook shown here is called the Codex Forster III and was one of many in a collection of his notebooks. This page shows da Vinci's pulley system which involves multiple gears and cogs. His notebook was filled with hundreds of depictions like this one as well as multiple sketches of artwork. Even engineers who lived hundreds of years ago used simple things, like project planning notebooks, to accomplish great tasks.

Plan a Hydraulic Robot / Project Planning

Plan a Hydraulic Robot / Project Planning

ACTIVITY

Plan a Hydraulic Robot / **Project Planning**

Plan a Hydraulic Robot / **Project Planning**

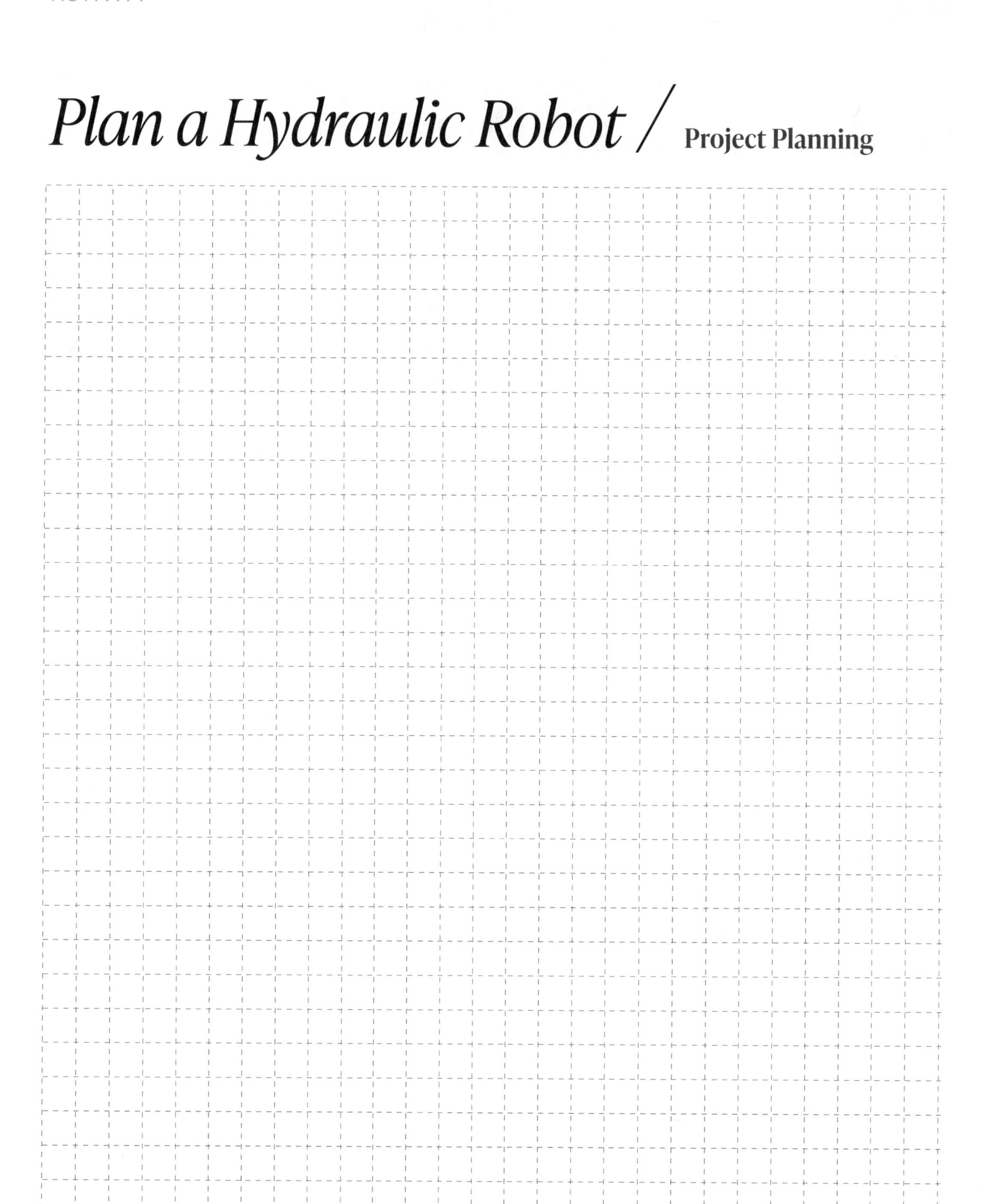

Plan a Hydraulic Robot / Project Planning

HYDRAULIC EXCAVATORS

LESSONS

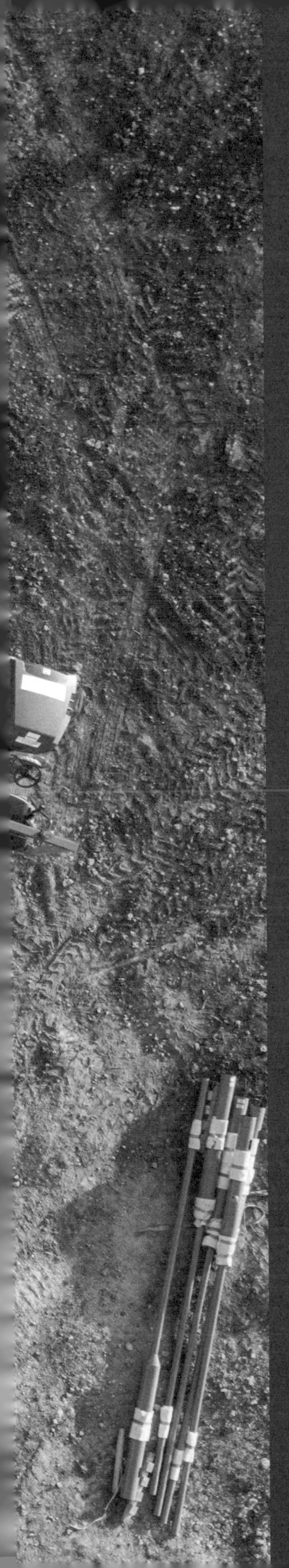

Hydraulic Robot: Construction

ROADMAP

- Understand how to follow a project plan.
- Learn how to work as a team and how to collaborate.
- Be able to work from orthographic projections and isometric sketches to build a structure.
- Be able to modify a plan when challenges occur.

Build a Hydraulic Robot / Construction

For this experiment, you will need the following materials:

- 60 simple craft sticks or 50 colorful craft sticks
- 30 wooden cubes
- 15 wooden cubes with holes
- 1 dowel
- 4 syringes
- 2 syringe caps
- 3 feet of tubing
- 2 small zip ties
- Hot glue gun and sticks
- Masking or painter's tape
- Scissors
- Balsa wood cutter
- Ruler & Pencil

Notes: This is not an instruction manual for your robot. It is only an example of how to build the robot with a specific set of materials. It is your job to apply the knowledge you have gained throughout this book to create your own hydraulic robot with the materials you are provided. You may use the instruction manual as a reference, but your design must be your own.

Step 1	Instructions
FIGURE 13.1	Before you begin, put on your safety glasses! With the same partner from the previous lesson, begin construction of your hydro robot according to the design which you planned together. The pieces you will use to build your hydraulic robot are shown in Figure 13.1.

Step 2	Instructions
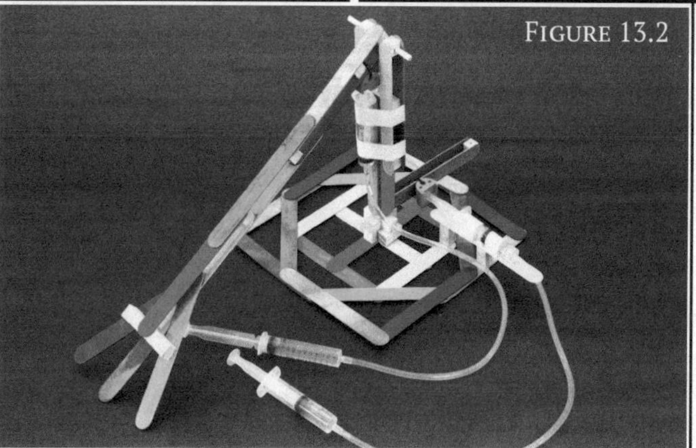 FIGURE 13.2	Your completed hydraulic robot might be similar to the example given in Figure 13.2 if you are using colorful craft sticks. Use this design as an inspiration. As you begin your work, make sure that any pieces that do not fit back into your kit are labeled with your name.

Build a Hydraulic Robot / Construction

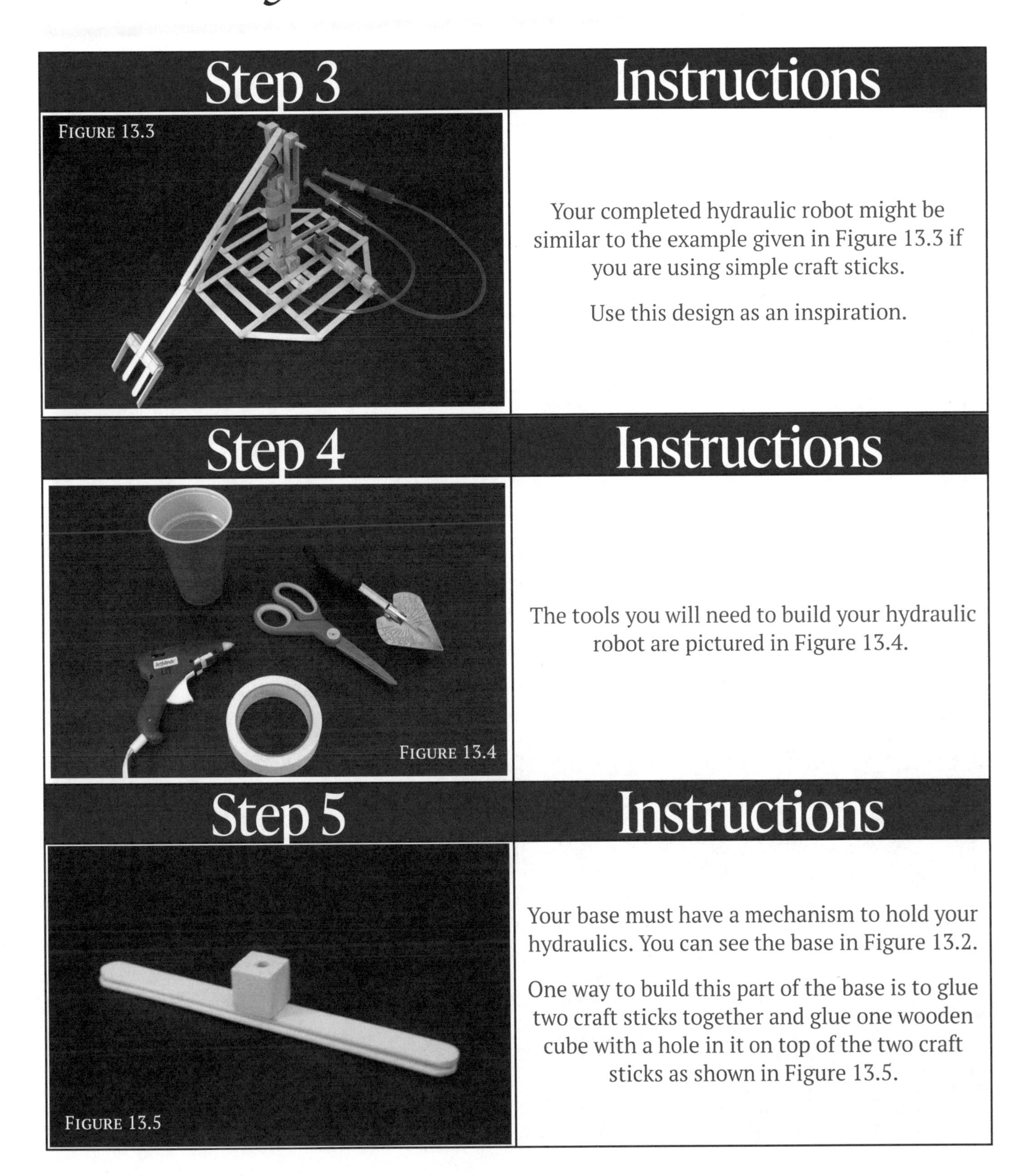

Step 3	Instructions
FIGURE 13.3	Your completed hydraulic robot might be similar to the example given in Figure 13.3 if you are using simple craft sticks. Use this design as an inspiration.

Step 4	Instructions
FIGURE 13.4	The tools you will need to build your hydraulic robot are pictured in Figure 13.4.

Step 5	Instructions
FIGURE 13.5	Your base must have a mechanism to hold your hydraulics. You can see the base in Figure 13.2. One way to build this part of the base is to glue two craft sticks together and glue one wooden cube with a hole in it on top of the two craft sticks as shown in Figure 13.5.

Build a Hydraulic Robot / Construction

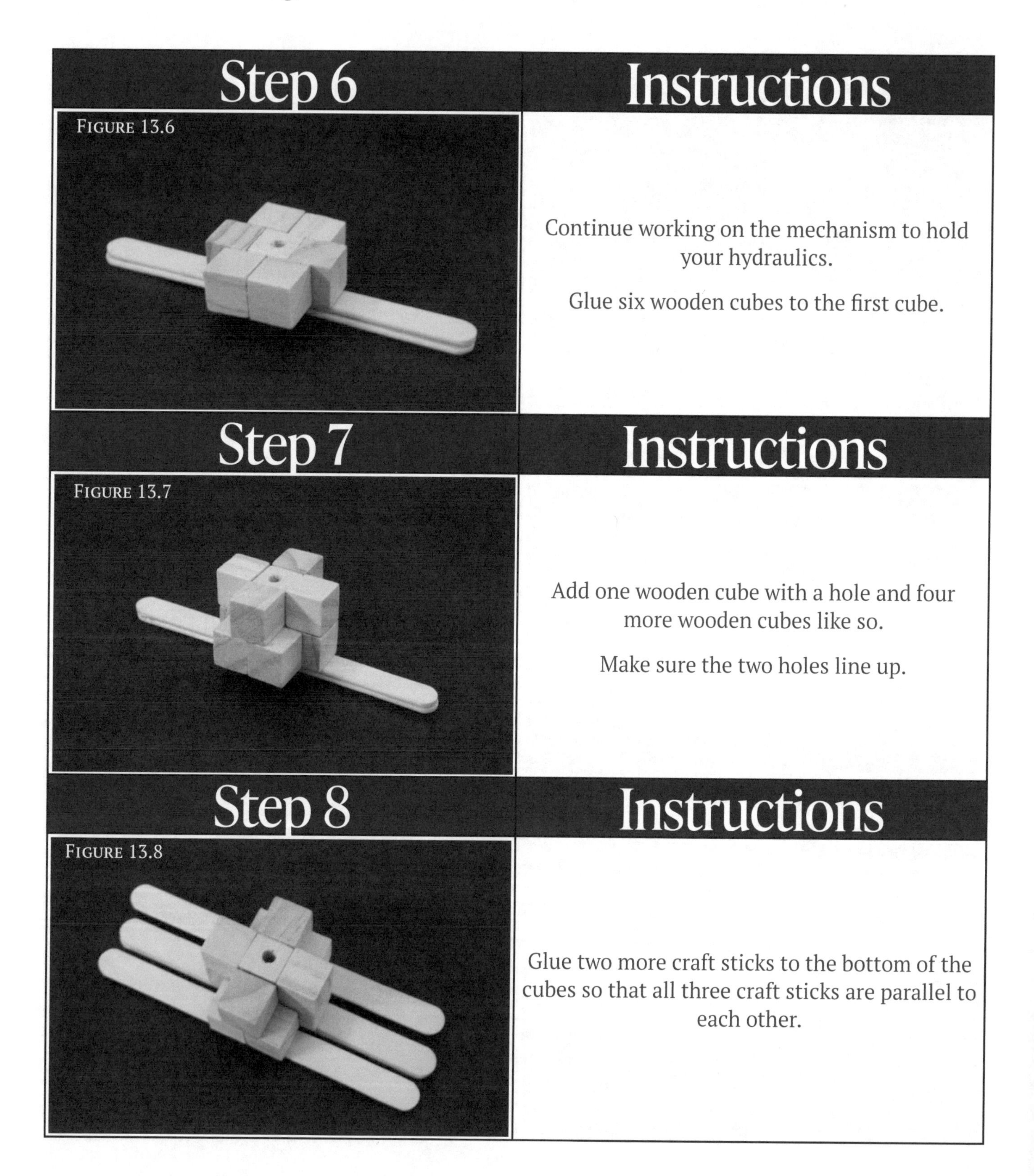

Step 6	Instructions
Figure 13.6	Continue working on the mechanism to hold your hydraulics. Glue six wooden cubes to the first cube.

Step 7	Instructions
Figure 13.7	Add one wooden cube with a hole and four more wooden cubes like so. Make sure the two holes line up.

Step 8	Instructions
Figure 13.8	Glue two more craft sticks to the bottom of the cubes so that all three craft sticks are parallel to each other.

Build a Hydraulic Robot / Construction

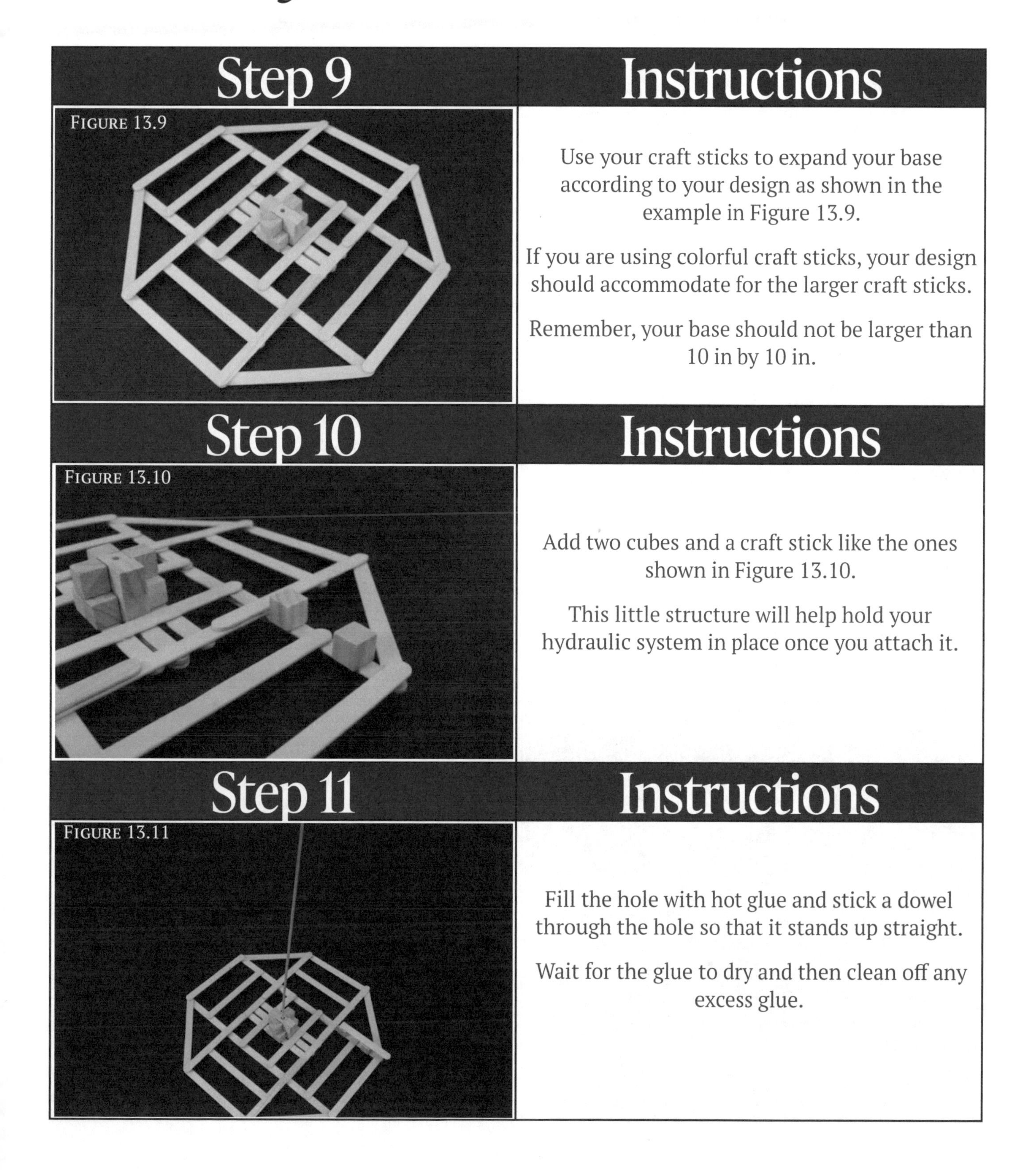

Step 9	Instructions
FIGURE 13.9	Use your craft sticks to expand your base according to your design as shown in the example in Figure 13.9. If you are using colorful craft sticks, your design should accommodate for the larger craft sticks. Remember, your base should not be larger than 10 in by 10 in.

Step 10	Instructions
FIGURE 13.10	Add two cubes and a craft stick like the ones shown in Figure 13.10. This little structure will help hold your hydraulic system in place once you attach it.

Step 11	Instructions
FIGURE 13.11	Fill the hole with hot glue and stick a dowel through the hole so that it stands up straight. Wait for the glue to dry and then clean off any excess glue.

Build a Hydraulic Robot / Construction

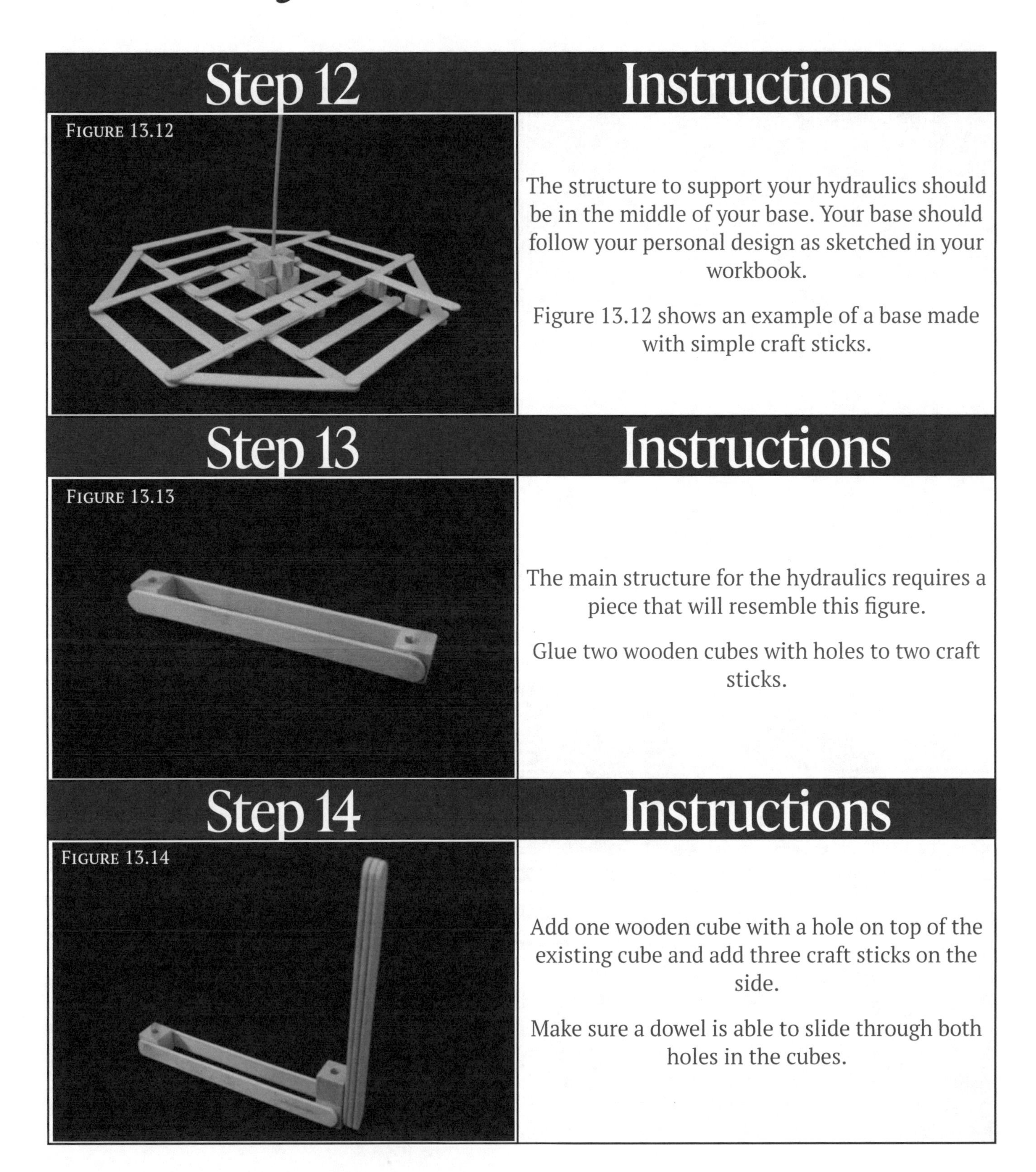

Step 12

FIGURE 13.12

Instructions

The structure to support your hydraulics should be in the middle of your base. Your base should follow your personal design as sketched in your workbook.

Figure 13.12 shows an example of a base made with simple craft sticks.

Step 13

FIGURE 13.13

Instructions

The main structure for the hydraulics requires a piece that will resemble this figure.

Glue two wooden cubes with holes to two craft sticks.

Step 14

FIGURE 13.14

Instructions

Add one wooden cube with a hole on top of the existing cube and add three craft sticks on the side.

Make sure a dowel is able to slide through both holes in the cubes.

Build a Hydraulic Robot / Construction

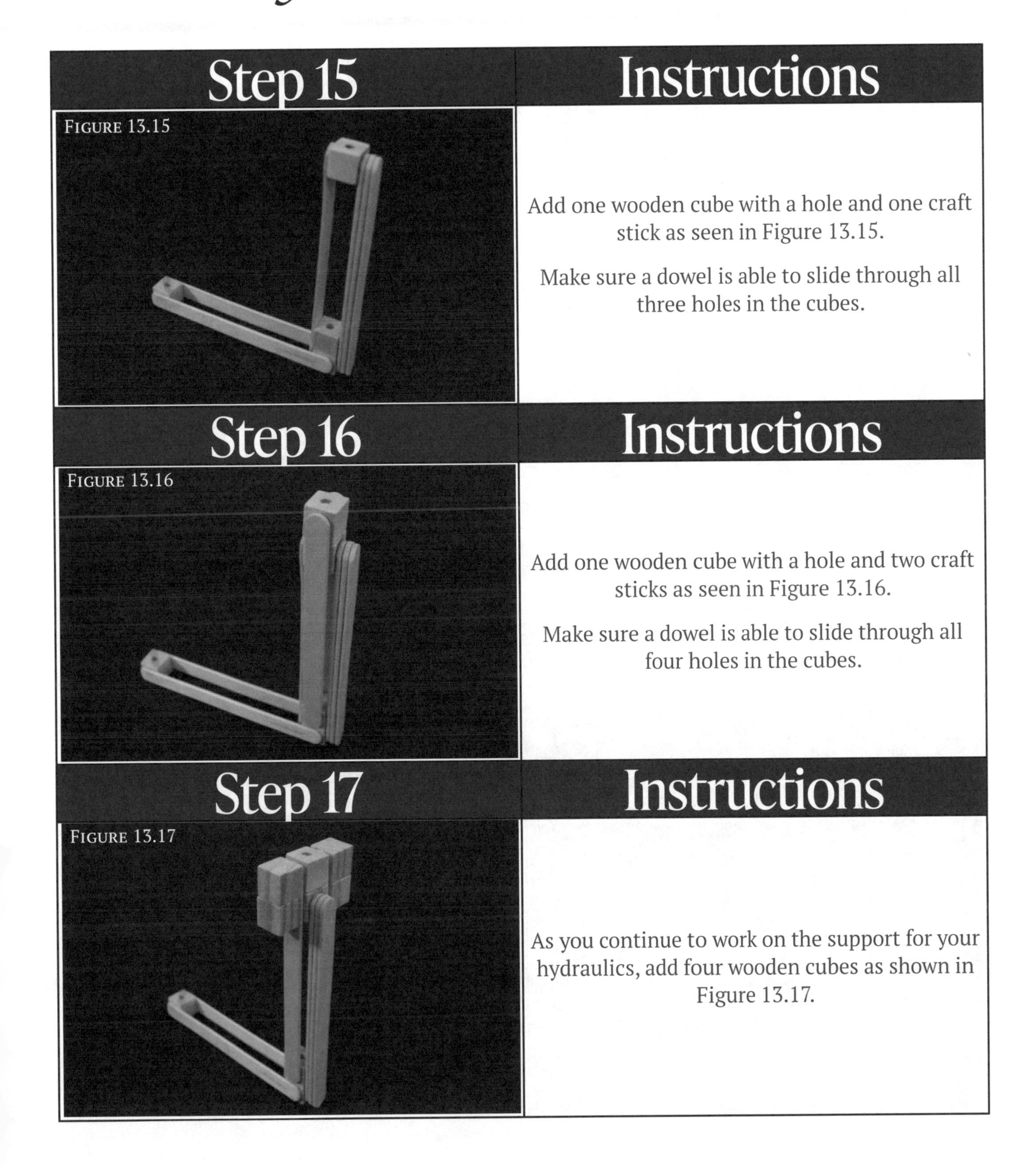

Step 15	Instructions
FIGURE 13.15	Add one wooden cube with a hole and one craft stick as seen in Figure 13.15. Make sure a dowel is able to slide through all three holes in the cubes.
Step 16	**Instructions**
FIGURE 13.16	Add one wooden cube with a hole and two craft sticks as seen in Figure 13.16. Make sure a dowel is able to slide through all four holes in the cubes.
Step 17	**Instructions**
FIGURE 13.17	As you continue to work on the support for your hydraulics, add four wooden cubes as shown in Figure 13.17.

Build a Hydraulic Robot / Construction

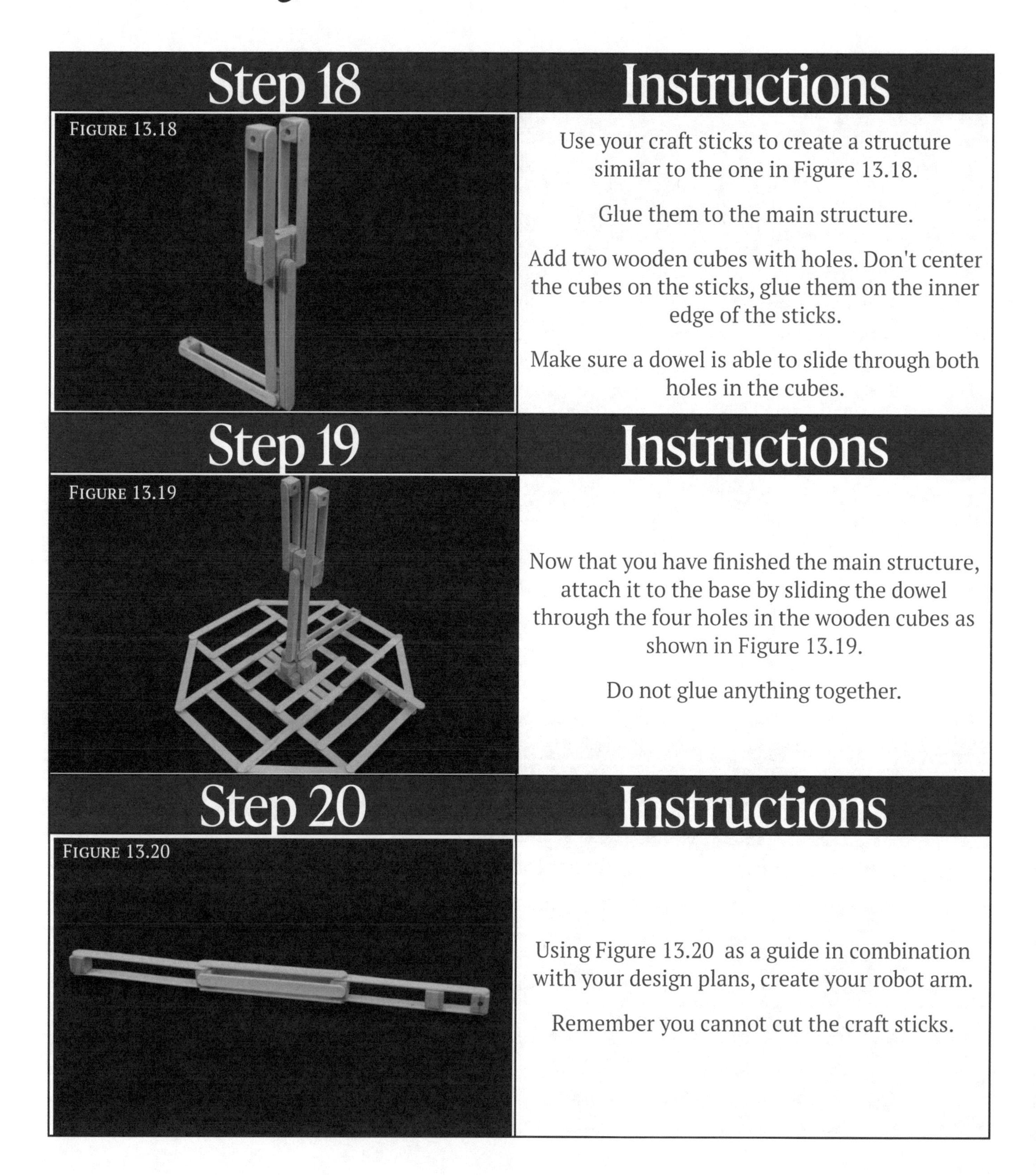

Step 18

FIGURE 13.18

Instructions

Use your craft sticks to create a structure similar to the one in Figure 13.18.

Glue them to the main structure.

Add two wooden cubes with holes. Don't center the cubes on the sticks, glue them on the inner edge of the sticks.

Make sure a dowel is able to slide through both holes in the cubes.

Step 19

FIGURE 13.19

Instructions

Now that you have finished the main structure, attach it to the base by sliding the dowel through the four holes in the wooden cubes as shown in Figure 13.19.

Do not glue anything together.

Step 20

FIGURE 13.20

Instructions

Using Figure 13.20 as a guide in combination with your design plans, create your robot arm.

Remember you cannot cut the craft sticks.

Build a Hydraulic Robot / Construction

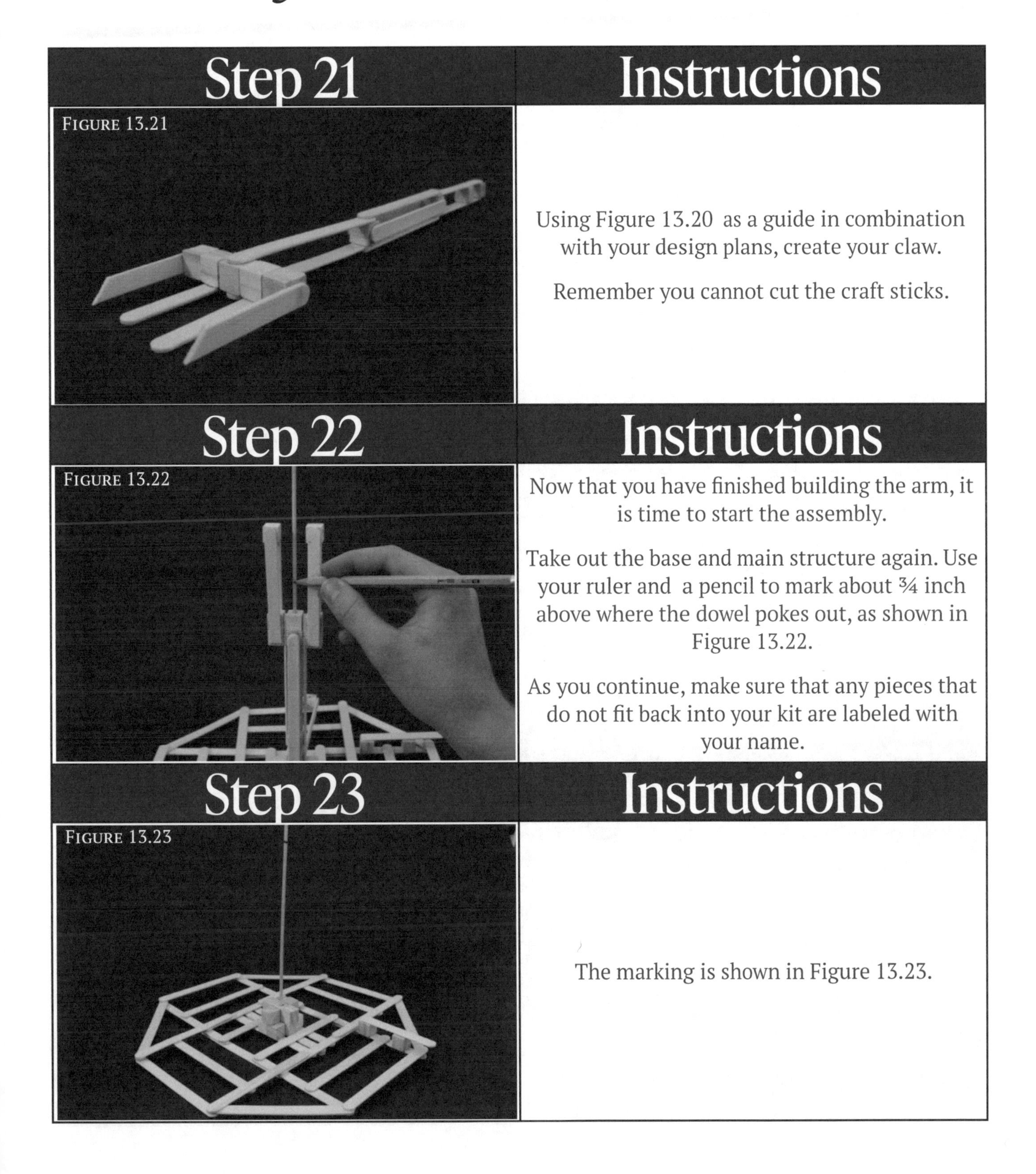

Step 21

FIGURE 13.21

Instructions

Using Figure 13.20 as a guide in combination with your design plans, create your claw.

Remember you cannot cut the craft sticks.

Step 22

FIGURE 13.22

Instructions

Now that you have finished building the arm, it is time to start the assembly.

Take out the base and main structure again. Use your ruler and a pencil to mark about ¾ inch above where the dowel pokes out, as shown in Figure 13.22.

As you continue, make sure that any pieces that do not fit back into your kit are labeled with your name.

Step 23

FIGURE 13.23

Instructions

The marking is shown in Figure 13.23.

Build a Hydraulic Robot / Construction

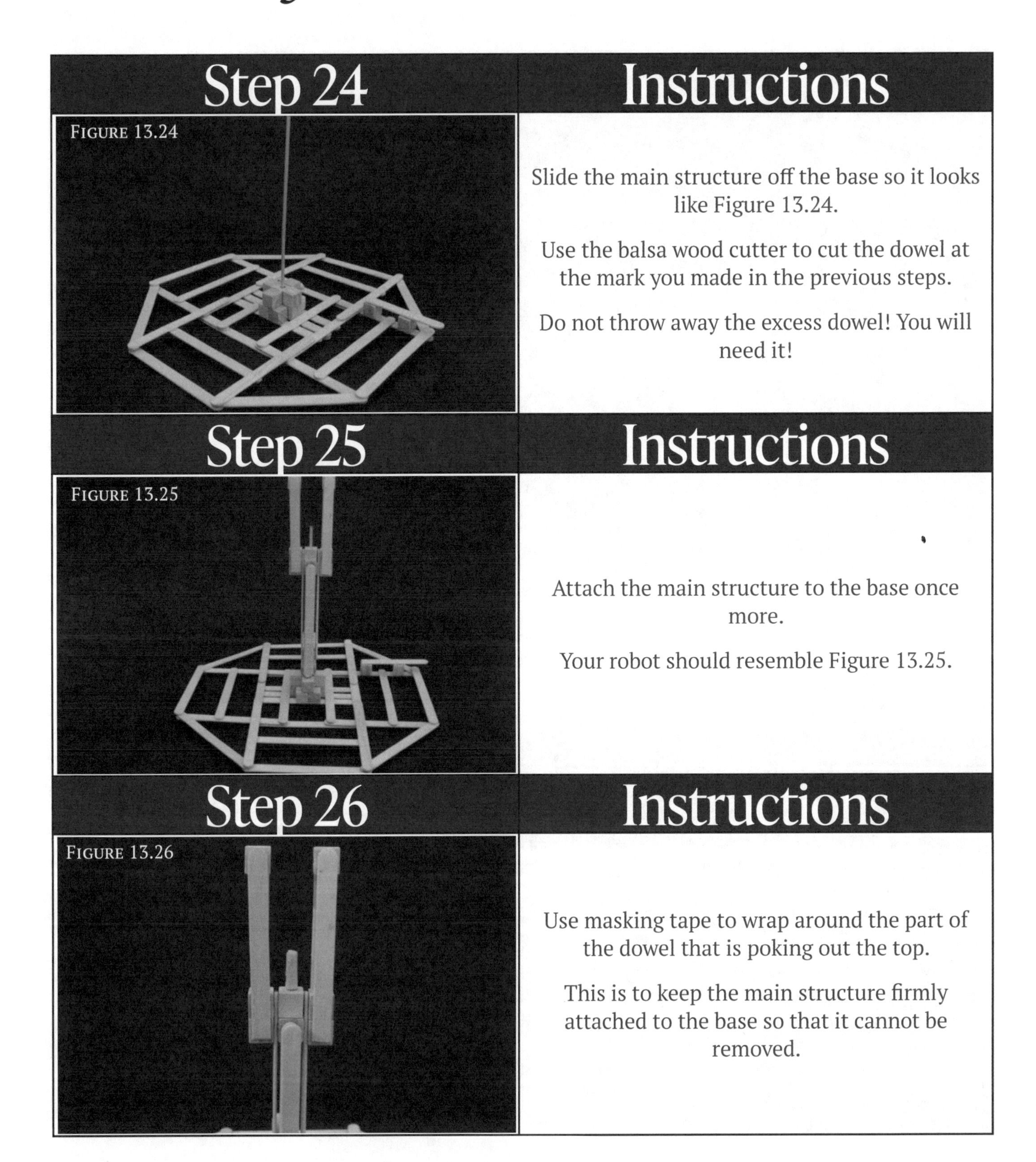

Step 24	Instructions
Figure 13.24	Slide the main structure off the base so it looks like Figure 13.24. Use the balsa wood cutter to cut the dowel at the mark you made in the previous steps. Do not throw away the excess dowel! You will need it!

Step 25	Instructions
Figure 13.25	Attach the main structure to the base once more. Your robot should resemble Figure 13.25.

Step 26	Instructions
Figure 13.26	Use masking tape to wrap around the part of the dowel that is poking out the top. This is to keep the main structure firmly attached to the base so that it cannot be removed.

Build a Hydraulic Robot / Construction

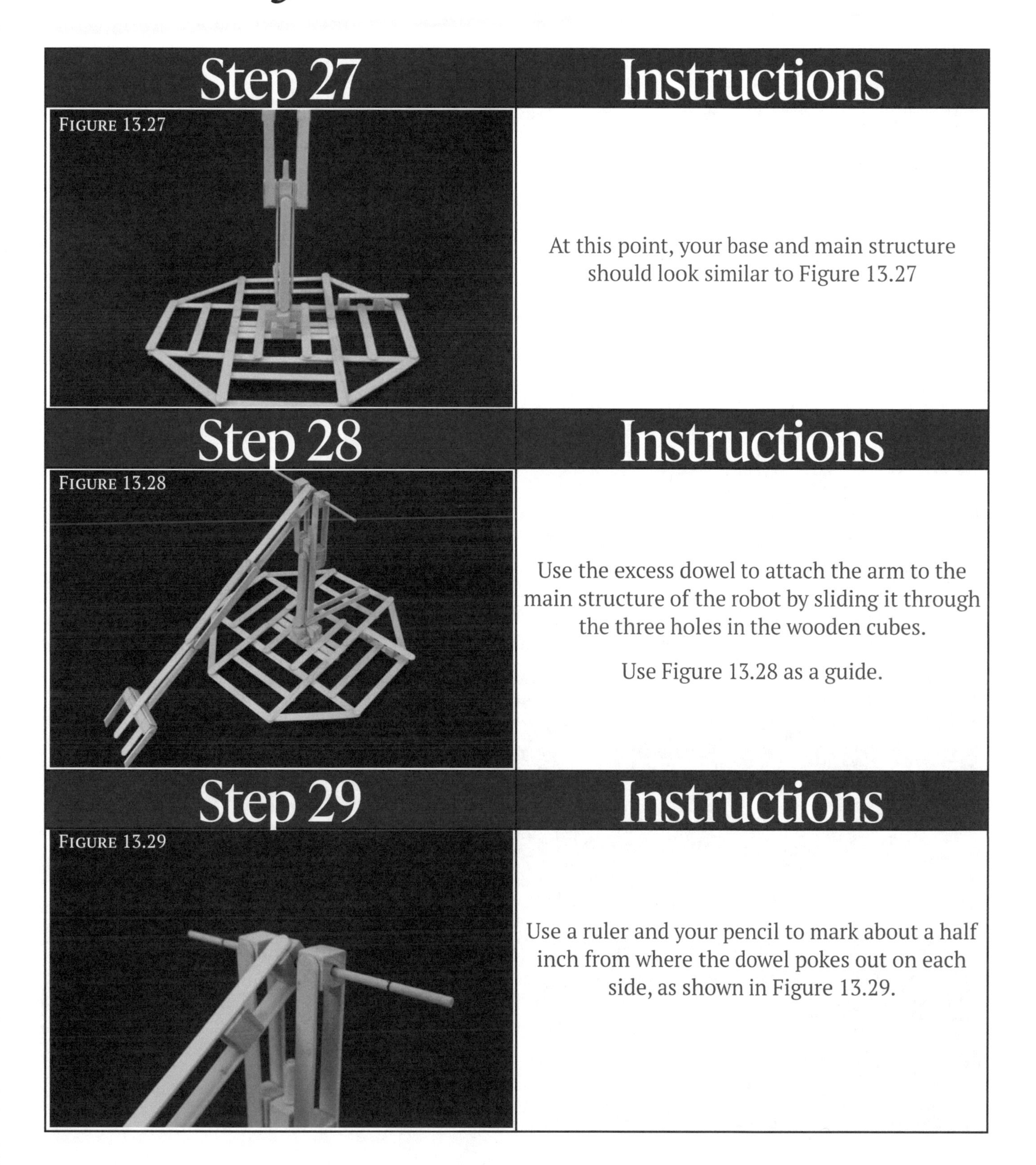

Step 27	Instructions
FIGURE 13.27	At this point, your base and main structure should look similar to Figure 13.27

Step 28	Instructions
FIGURE 13.28	Use the excess dowel to attach the arm to the main structure of the robot by sliding it through the three holes in the wooden cubes. Use Figure 13.28 as a guide.

Step 29	Instructions
FIGURE 13.29	Use a ruler and your pencil to mark about a half inch from where the dowel pokes out on each side, as shown in Figure 13.29.

Build a Hydraulic Robot / Construction

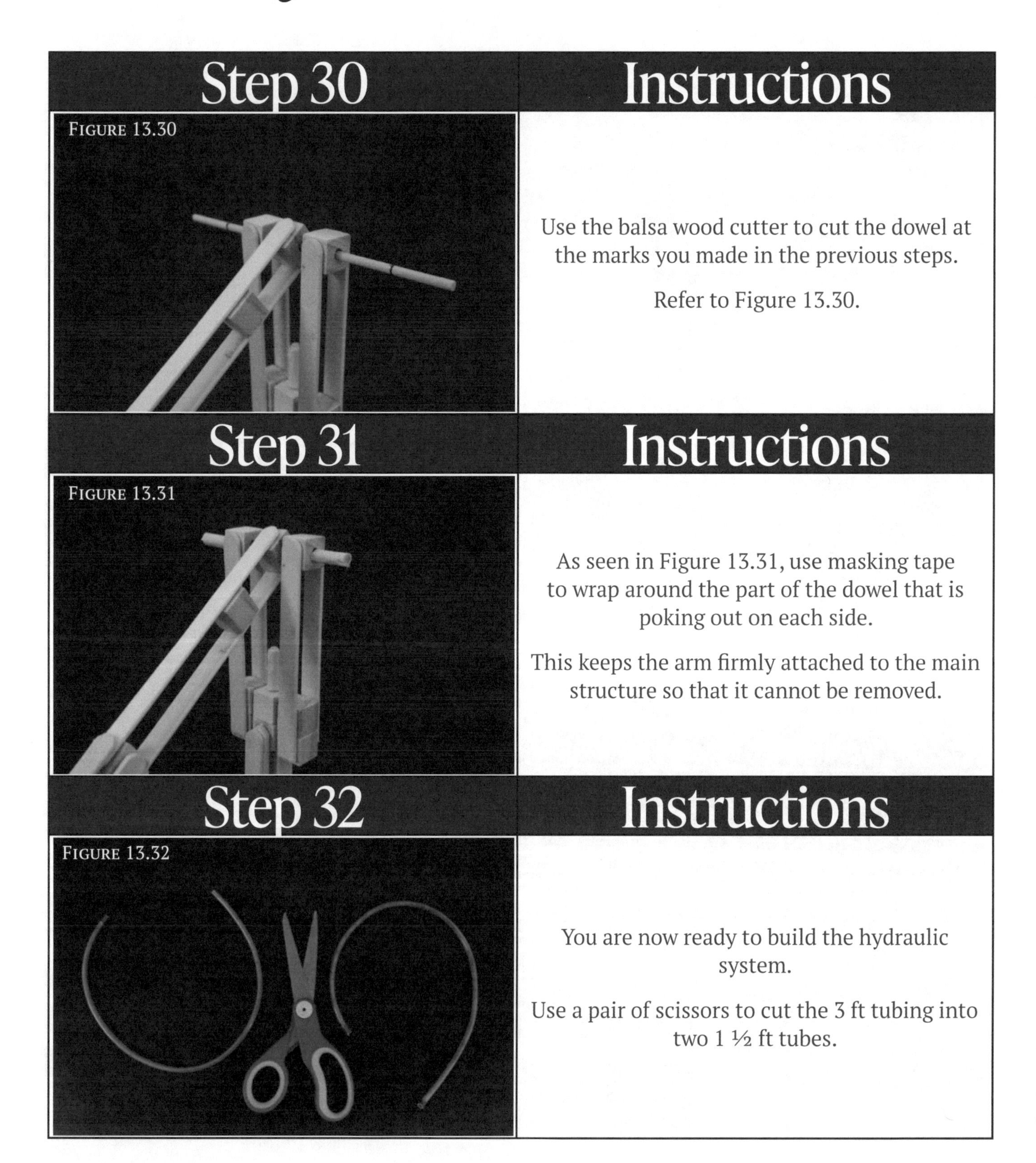

Step 30	Instructions
FIGURE 13.30	Use the balsa wood cutter to cut the dowel at the marks you made in the previous steps. Refer to Figure 13.30.
Step 31	**Instructions**
FIGURE 13.31	As seen in Figure 13.31, use masking tape to wrap around the part of the dowel that is poking out on each side. This keeps the arm firmly attached to the main structure so that it cannot be removed.
Step 32	**Instructions**
FIGURE 13.32	You are now ready to build the hydraulic system. Use a pair of scissors to cut the 3 ft tubing into two 1 ½ ft tubes.

Build a Hydraulic Robot / Construction

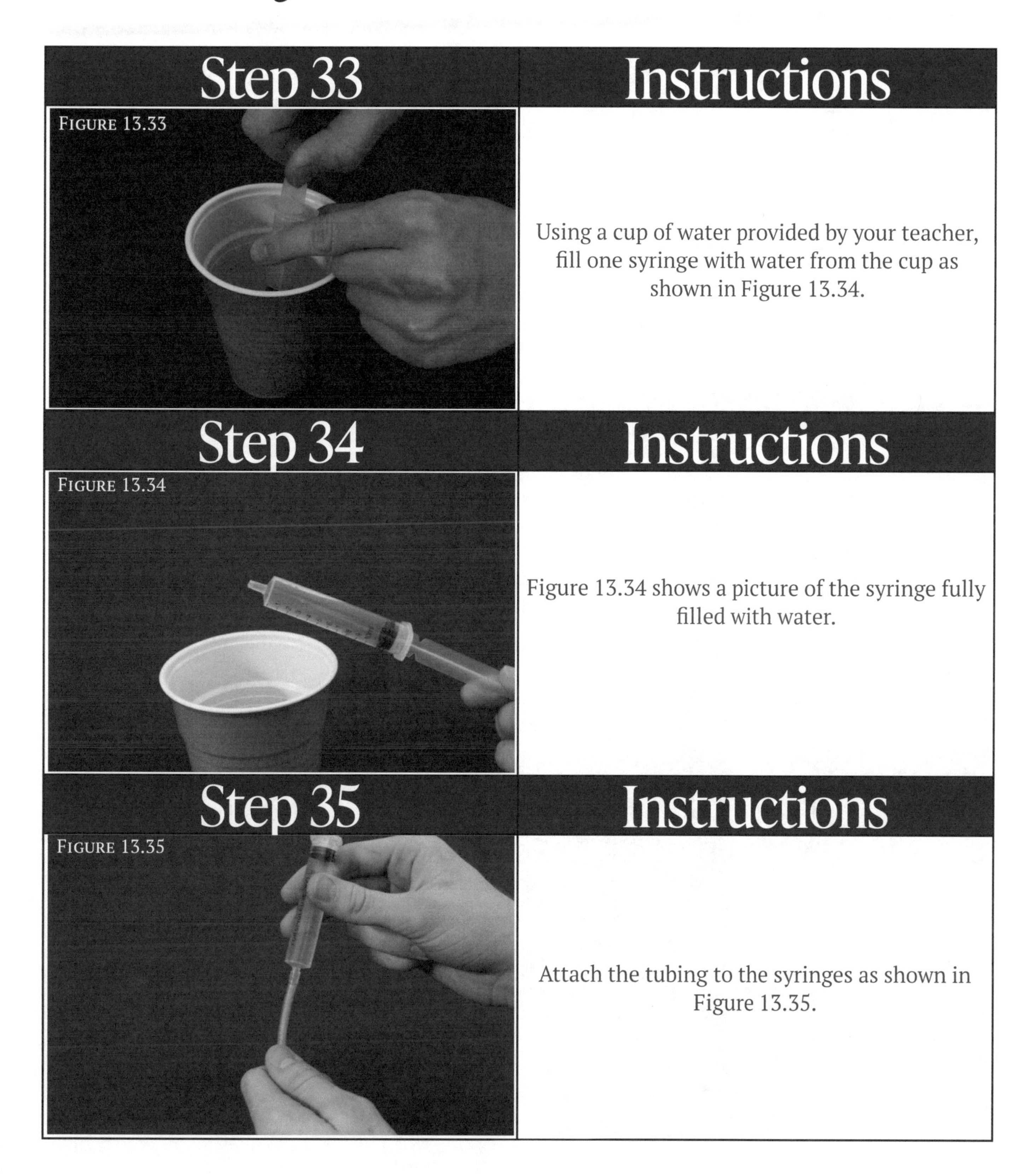

Step 33	Instructions
FIGURE 13.33	Using a cup of water provided by your teacher, fill one syringe with water from the cup as shown in Figure 13.34.

Step 34	Instructions
FIGURE 13.34	Figure 13.34 shows a picture of the syringe fully filled with water.

Step 35	Instructions
FIGURE 13.35	Attach the tubing to the syringes as shown in Figure 13.35.

Build a Hydraulic Robot / Construction

Step 36	Instructions
FIGURE 13.36	Empty the water from the syringe through the tube and back into the cup as shown in Figure 13.36.
Step 37	**Instructions**
FIGURE 13.37	Fill the syringe with water through the tube.
Step 38	**Instructions**
FIGURE 13.38	Attach another syringe to the opposite end of the tube as shown in Figure 13.38.

Build a Hydraulic Robot / Construction

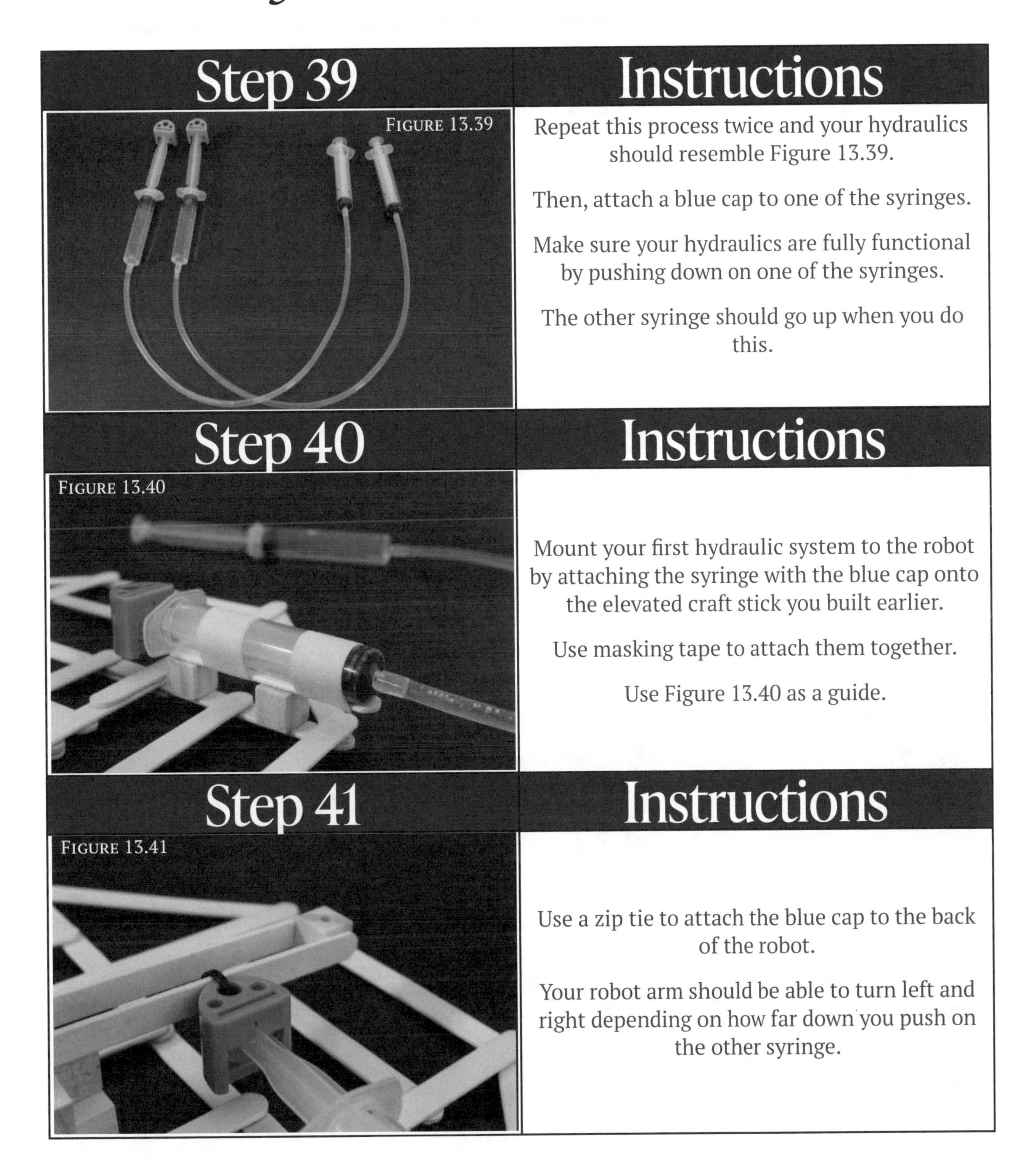

Step 39

FIGURE 13.39

Instructions

Repeat this process twice and your hydraulics should resemble Figure 13.39.

Then, attach a blue cap to one of the syringes.

Make sure your hydraulics are fully functional by pushing down on one of the syringes.

The other syringe should go up when you do this.

Step 40

FIGURE 13.40

Instructions

Mount your first hydraulic system to the robot by attaching the syringe with the blue cap onto the elevated craft stick you built earlier.

Use masking tape to attach them together.

Use Figure 13.40 as a guide.

Step 41

FIGURE 13.41

Instructions

Use a zip tie to attach the blue cap to the back of the robot.

Your robot arm should be able to turn left and right depending on how far down you push on the other syringe.

Build a Hydraulic Robot / Construction

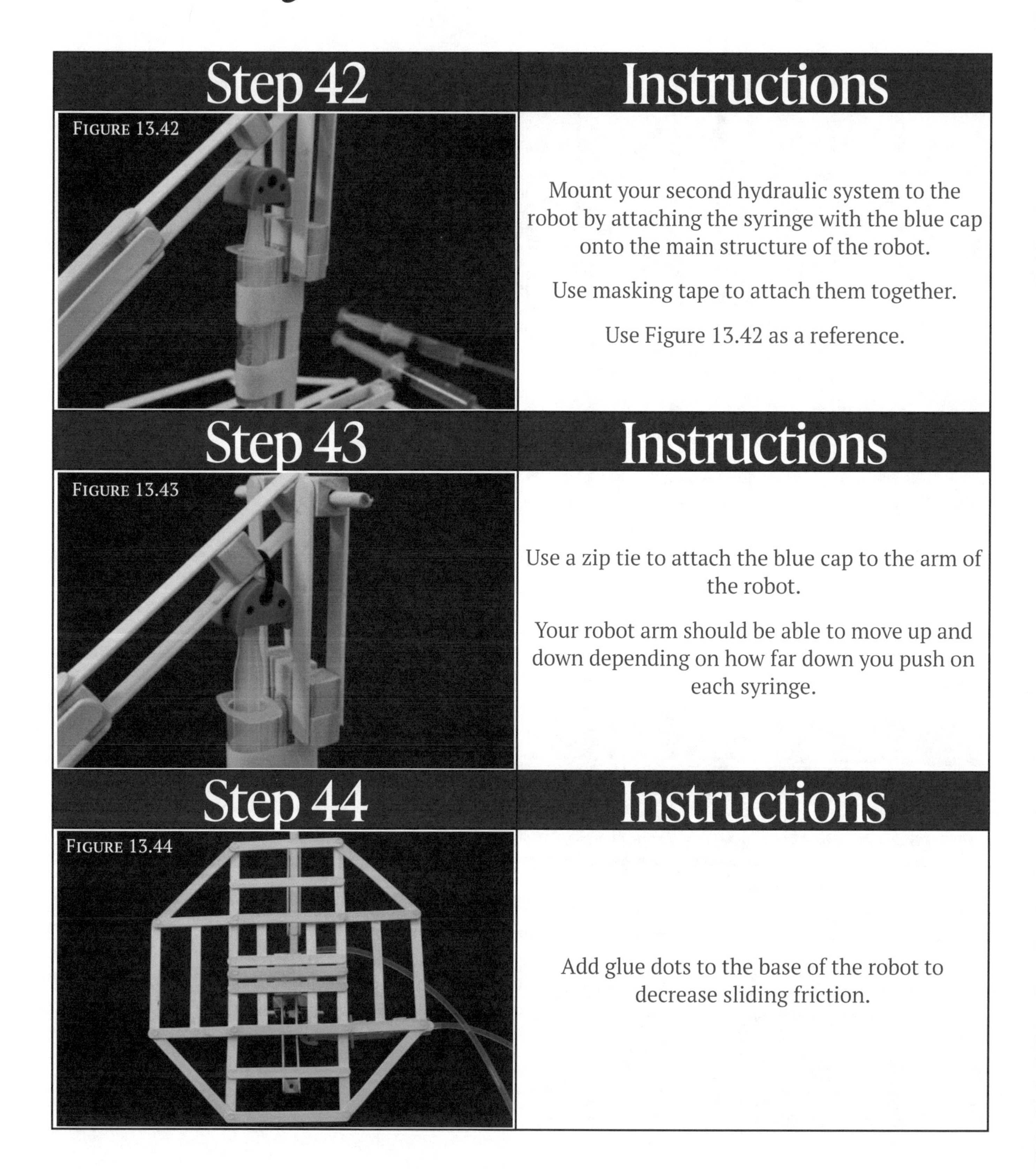

Step 42	Instructions
FIGURE 13.42	Mount your second hydraulic system to the robot by attaching the syringe with the blue cap onto the main structure of the robot. Use masking tape to attach them together. Use Figure 13.42 as a reference.
Step 43	**Instructions**
FIGURE 13.43	Use a zip tie to attach the blue cap to the arm of the robot. Your robot arm should be able to move up and down depending on how far down you push on each syringe.
Step 44	**Instructions**
FIGURE 13.44	Add glue dots to the base of the robot to decrease sliding friction.

Build a Hydraulic Robot / Construction

Step 45

FIGURE 13.45

Instructions

Once completed, your robot should be able to turn left and right using one syringe, and the arm should be able to move up and down using the other syringe.

You have officially completed your hydraulic robot!

Troubleshooting

If you are having trouble with the structure of your robot not remaining intact, or with the functionality of your robot, here are some steps to take to improve your build.

1. Make sure your build accurately represents what you have planned in your notebook, including all measurements and the placement of all craft sticks and cubes.
2. If you need to make changes and you have materials available, plan your modifications carefully.
3. If you need to detach a misplaced piece, make sure to detach the piece carefully without detaching or breaking adjacent pieces.
4. Make sure your craft sticks are positioned flat against each other and do not impede moving parts.
5. Make sure pieces attached to dowels can rotate freely.
6. Adjust tightness of zipties if necessary.

HYDRAULIC EXCAVATOR

Photo by Tobias Kleeb

LESSON

Hydraulic Robot: Testing & Competition

ROADMAP

- Understand the importance of testing a completed project.
- Practice analyzing what went wrong and what could have gone better.
- Learn to re-design for a better outcome the next time.

Hydraulic Robot: Testing & Competition

What is it and why is it important?

The hydraulic robot is a competition robot, but before you compete with other robots you and your partner first need to test it. You will need to record the results of these tests. Testing is an important part of the engineering design cycle, it allows you to find problems before they cause accidents. It is better to test an invention beyond its intended use to ensure it is strong enough to maintain its functions. It is better to have your invention break in a controlled environment rather than when it is in the process of being used. When testing, data collection is critically important. Your data should be detailed, and clear enough that others can understand your observations and data and what they mean.

How does it work?

In this lesson you will conduct several tests on your robot. These tests will be simple functions that the robot will use when competing. The tests include testing how well the arm pivots on the base, how well the arm moves up and down at different speeds, how the balance of your robot is when the arm is at different heights and pivoted in different directions on your base.

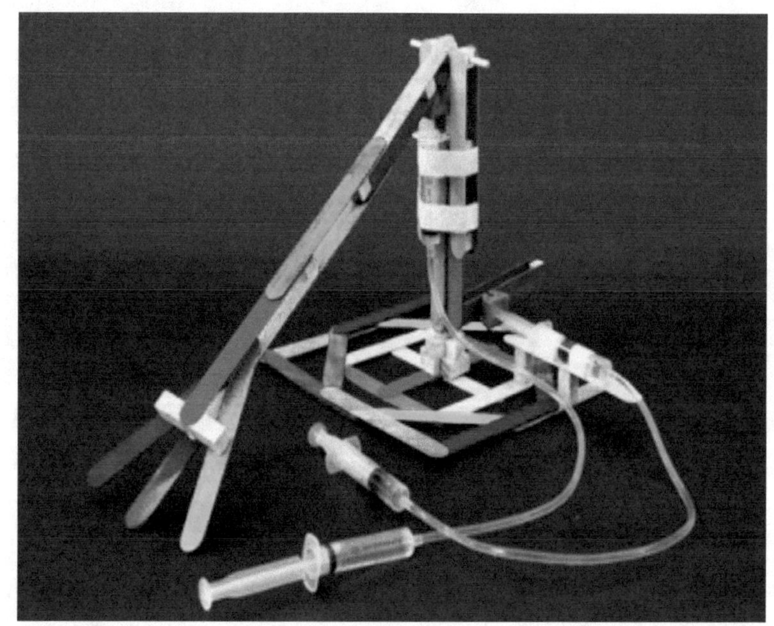

FIGURE 15.1
AN EXAMPLE OF A HYDRAULIC ROBOT

After you and your partner have completed all tests and recorded your data, you can repair any broken parts or move on to the competition phase. If you need to repair any broken parts go back to the previous lesson on construction of the robot. If no repairs are necessary, then you can begin competing with your robot.

Test a Hydraulic Robot / Testing & Competition

Now it is time to work with your partner to test your hydraulic robot.

Supplies needed are painters tape, scissors and a ruler.

In each section below, there is a question that you will use to test your robot.

Tests are divided into 4 categories: Hydraulics, Arm, Balance and Efficiency

1. Record all observations in the provided space.
2. Perform each test three times and record the results for each test.
3. Explain how you performed each test.

Test Question	***Student Observation***
Hydraulics 1. Are the hydraulics for rotating the assembly operating smoothly or are they stiff? 2. Are the hydraulics for rotating the base using their full range of motion, or are they stopping before full extension and contraction?	

ACTIVITY

Test a Hydraulic Robot / Testing & Competition

Test Question	*Student Observation*
Arm Function 1. Are the hydraulics for moving the arm up and down operating smoothly or are they stiff? 2. Are the hydraulics for for moving the arm up and down using their full range of motion, or are they stopping before full extension and contraction?	

Test Question	*Student Observation*
Arm Function 3. Starting when the arm is all the way up, how long does it take for the arm to fall all the way down in seconds?	

Test a Hydraulic Robot / Testing & Competition

Now it is time to test the robot's balance and stability in different positions.

We will do this by testing the robot's balance when the assembly is pivoted in different directions.

We will also position the arms at different heights.

1. When the test question states "Position 1", fully contract both of your hydraulics.
2. When the test question states "Position 2", fully extend both of your hydraulics.
3. Make sure not to change the robot's position during the course of each test.

Test Question	*Student Observation*
Stability: Robot Arm Down 4. Is the robot tipping over in any direction or is the robot stable all the way around when the robot is in: a. Position 1? b. Position 2?	

ACTIVITY

Test a Hydraulic Robot / Testing & Competition

Test Question	*Student Observation*
Stability: Robot Arm Half Way Up 5. Is the robot tipping over in any direction or is the robot stable all the way around when the robot is in: a. Position 1? b. Position 2?	

Test Question	*Student Observation*
Stability: Robot Arm Up 6. Is the robot tipping over in any direction or is the robot stable all the way around when the robot is in: a. Position 1? b. Position 2?	

Test a Hydraulic Robot / Testing & Competition

Test Question	Student Observation
Efficiency and Balance While Both Hydraulics are Being Used 7. When using both sets of hydraulics at the same time, are the hydraulics smooth or stiff?	

Test Question	Student Observation
Efficiency and Balance While Both Hydraulics are Being Used 8. When using both sets of hydraulics at the same time, does the robot lose balance or tip over?	

ACTIVITY

Test a Hydraulic Robot / Testing & Competition

Test Question	***Student Observation***
Overall Performance 9. Overall, How do you think your robot performed during the testing?	
Instructions	***Student Conclusions***
Review the performance of your robot during the 9 parts of the robot test. If your robot did not perform well, write down the problem that occurred and the possible solution. Before moving on to the competition, modify your robot to solve any problems discovered during testing.	

ACTIVITY

Robot Competition / Testing & Competition

Instructions: For the following lesson experiment, work with your partner to try out-performing other students' hydraulic robots in a competition!

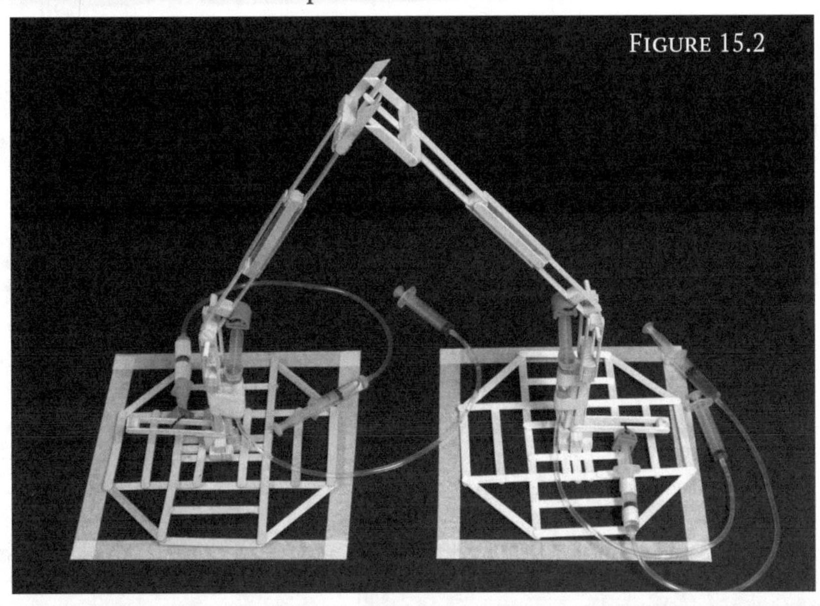

FIGURE 15.2

Step	Instructions	Photo
Step 1	• Create a square using painter's tape that is 10 in by 10 in. • Then measure 3 inches in front of the first square, and make another 10 in by 10 in square with painters tape. • Use Figure 15.3 as a reference.	FIGURE 15.3
Step 2	• Place a robot in each square and have them face each other. • Have each robot start with their arm all the way up like shown in Figure 15.4. • Have a third person count down and then begin the competition.	FIGURE 15.4

Robot Competition / Testing & Competition

Step	Instructions	Photo
Step 3	How to win a competition: • If you push your opponent more than halfway outside the box, you win. Refer to figure 15.5. • If you flip your opponent over, you win. Refer to Figure 15.6. • In the case of a draw, restart the competition.	FIGURE 15.5
Step 4	Recording results: • Conduct at least 3 competitions with 3 different robots. • Based on what happens in these 3 competitions, fill out the following table.	FIGURE 15.6

Robot competitions are very popular and robotics teams offer opportunities to learn many skills, including robot construction, programming, leadership, and collaboration. Check out your Thales Academy campus and see if your school has a team that you can join!

ACTIVITY

Robot Competition / Testing & Competition

Instructions: Based on what happens in these 3 competitions, fill out the table.

Competition	Opponent Name	Win / Lose	Way of Win / Lose	# of Draws	Observations
Competition #1	______	______	______	______	______
Competition #2	______	______	______	______	______
Competition #3	______	______	______	______	______

Photography Citations

mino21/Adobe Stock .... *Page 6-7*

NASA / Unsplash license .... *Page 8-9*

CaptiveAire Systems .... *Page 10-11*

Graphic / Adobe Stock #489361967 ..... *Page 12*

Rawf8 / Adobe Stock .... *Page 13*

Josh Stewart / Unsplash license ..... *Page 16-17*

Winston Brady .... *Figure 2.1*

Graphic / Adobe Stock #10056459 ..... *Page 18*

Glenn Carstens-Peters / Unsplash license . *Page 19*

Thomas Alva Edison by Louis Bachrach, 1922 .. *Page 20*
https://en.wikipedia.org/wiki/Thomas_Edison#/media/File:Thomas_Edison2.jpg

Graphic / Adobe Stock #124864689 ..... *Page 21*

Graphic / Adobe Stock #208537836 ..... *Page 25*

Hannes Richter / Unsplash license .... *Page 30-31*

Florida Center for Instructional Technology .. *Figure 3.1*

NASA/JPL/Corby Waste, 2014 .... *Page 10-11*
http://www.vitalstatistics.info/uploads/mars%20climate%20orbiter.jpg

Tommao Wang / Unsplash license .... *Page 40-41*

attaphong / Vector Stock .... *Figure 4.1*

CaptiveAire Systems .... *Figure 4.2*

Andrew Baird .... *Figure 4.3*

Andrew Baird .... *Figure 4.4*

Zhang Zeduan, 1085 .... *Page 10-11*
https://en.wikipedia.org/wiki/Parallel_projection#/media/File:Along_the_River_During_the_Qingming_Festival_(detail_of_original).jpg

Bush 'o' graphy / Unsplash license .... *Page 50-51*

BaMic Illustrations / Adobe Stock .... *Figure 5.1*

Sebastian Le Clerc, 1684 .... *Page 52*
https://en.wikipedia.org/wiki/Vitruvius#/media/File:Vitruvius.jpg

BaMic Illustrations / Adobe Stock .... *Figure 5.2*

Andrew Baird .... *Figure 5.4*

Cristina Gottardi / Unsplash license ..... *Page 60-61*

Brovarky / Adobe Stock .... *Figure 6.1*

Comauthor / Adobe Stock .... *Figure 6.2*

supakitmod / Adobe Stock .... *Figure 6.3*

Funtay / Adobe Stock .... *Figure 6.4*

Scott Brody, 2017 .... *Figure 6.5*
https://en.wikipedia.org/wiki/Tape_measure#/media/File:Dual_Scale_Stanley_Tape_Measure.jpg

Gresei / Adobe Stock .... *Figure 6.6*

Alpar / Adobe Stock .... *Figure 6.7*

Charles Givens / Unsplash license .... *Page 68-69*

Ivan Burchak / Adobe Stock .... *Figure 7.1*

demidoff / Adobe Stock .... *Figure 7.2*

Ekler / Adobe Stock .... *Figure 7.3*

Ekler / Adobe Stock .... *Figure 7.4*

Ekler / Adobe Stock .... *Figure 7.5*

No Author Provided, 14 April 2006 .... *Page 75*
https://en.wikipedia.org/wiki/Anubis#/media/File:Egypt_dauingevekten.jpg

Pavol Svantner / Unsplash license .... *Page 76-77*

Unknown / Victoria and Albert Museum ...... *Page 78*

zzzdim / Adobe Stock .... *Page 78*

Graphic / Adobe Stock #64358978 ..... *Page 25*

Photography Citations

Chaosamran_Studi / Adobe Stock • • • • • *Page 88*

Dominik Vanyi / Unsplash license • • • • • *Page 90-91*

Graphic / Adobe Stock #485017055 • • • • • *Page 94*

Winston Brady • • • • • *Figure 10.3*

An Qi Wang / Adobe Stock • • • • • *Page 96*

Cyberphysics, 2024 • • • • • *Figure 10.4*
https://www.cyberphysics.co.uk/Q&A/KS3/moments/A2.html

Eduardo Estellez / Adobe Stock • • • • • *Page 99*

Fly&I / Unsplash license • • • • • *Page 100-101*

photo 5000 / Adobe Stock • • • • • *Figure 11.1*

VectorMine / Adobe Stock • • • • • *Figure 11.2*

Grispb / Adobe Stock • • • • • *Figure 11.4*

Creativa Images / Adobe Stock • • • • • *Figure 11.5*

Philippe / Adobe Stock • • • • • *Figure 11.6*

fesenko / Adobe Stock • • • • • *Figure 11.7*

RYosha / iStock • • • • • *Page 106*

Unbekannt, 1750 • • • • • *Page 107*
https://en.wikipedia.org/wiki/Daniel_Bernoulli#/media/File:Portr%C3%A4t_des_Daniel_Bernoulli_-_edit1.jpg

Austin Williams • • • • • *Page 109*

MaxSafaniuk / Adobe Stock • • • • • *Page 110-111*

Andrew Baird • • • • • *Figure 12.2*

Andrew Baird • • • • • *Figure 12.3*

Andrew Baird • • • • • *Figure 12.4*

Andrew Baird • • • • • *Figure 12.5*

Leonardo da Vinci, 1400-1500 • • • • • *Page 118*
https://en.wikipedia.org/wiki/Science_and_inventions_of_Leonardo_da_Vinci#/media/File:Leonardo_machines.JPG

ink drop / Adobe Stock • • • • • *Page 124-125*

Tobias Kleeb / Unsplash license • • • • • *Page 142-143*

Batteries Not Included • • • • • *Page 152*

Andrew Baird & Nolan Sawyer • • • • • *Figure 4.3*

Andrew Baird & Nolan Sawyer • • • • • *Figure 4.4*

Andrew Baird & Nolan Sawyer • • • • • *Figure 4.5*

Andrew Baird & Nolan Sawyer • • • • • *Figure 4.6*

Andrew Baird & Nolan Sawyer • • • • • *Figure 4.7*

Andrew Baird & Nolan Sawyer • • • • • *Figure 4.8*

Andrew Baird & Nolan Sawyer • • • • • *Figure 5.3*

Andrew Baird & Nolan Sawyer • • • • • *Figure 5.4*

Andrew Baird & Nolan Sawyer • • • • • *Figure 5.5*

Andrew Baird & Nolan Sawyer • • • • • *Figure 5.6*

Andrew Baird & Nolan Sawyer • • • • • *Figure 5.7*

Andrew Baird & Nolan Sawyer • • • • • *Figure 8.1*

Andrew Baird & Nolan Sawyer • • • • • *Figure 8.2*

Andrew Baird & Nolan Sawyer • • • • • *Figure 8.4*

Andrew Baird & Nolan Sawyer • • • • • *Figure 8.5*

Andrew Baird & Nolan Sawyer • • • • • *Figure 8.6*

Andrew Baird & Nolan Sawyer • • • • • *Figure 8.7*

Andrew Baird & Nolan Sawyer • • • • • *Figure 8.8*

Andrew Baird & Nolan Sawyer • • • • • *Figure 8.9*

Photography Citations

Andrew Baird & Nolan Sawyer .... *Figure 8.10*

Andrew Baird & Nolan Sawyer .... *Figure 8.11*

Andrew Baird & Nolan Sawyer .... *Figure 8.12*

Andrew Baird & Nolan Sawyer .... *Figure 8.13*

Andrew Baird & Nolan Sawyer .... *Figure 8.14*

Andrew Baird & Nolan Sawyer .... *Figure 8.15*

Andrew Baird & Nolan Sawyer .... *Figure 8.16*

Andrew Baird & Nolan Sawyer .... *Figure 8.17*

Andrew Baird & Nolan Sawyer .... *Figure 10.1*

Andrew Baird & Nolan Sawyer .... *Figure 10.2*

Andrew Baird & Nolan Sawyer .... *Figure 11.3*

Andrew Baird & Nolan Sawyer .... *Figure 11.8*

Andrew Baird & Nolan Sawyer .... *Figure 11.9*

Andrew Baird & Nolan Sawyer .... *Figure 12.1*

Andrew Baird & Nolan Sawyer .... *Figure 12.6*

Andrew Baird & Nolan Sawyer .... *Figure 12.7*

Andrew Baird & Nolan Sawyer .... *Figure 12.8*

Andrew Baird & Nolan Sawyer .... *Figure 13.1*

Andrew Baird & Nolan Sawyer .... *Figure 13.2*

Andrew Baird & Nolan Sawyer .... *Figure 13.3*

Andrew Baird & Nolan Sawyer .... *Figure 13.4*

Andrew Baird & Nolan Sawyer .... *Figure 13.5*

Andrew Baird & Nolan Sawyer .... *Figure 13.6*

Andrew Baird & Nolan Sawyer .... *Figure 13.7*

Andrew Baird & Nolan Sawyer .... *Figure 13.8*

Andrew Baird & Nolan Sawyer .... *Figure 13.9*

Andrew Baird & Nolan Sawyer .... *Figure 13.10*

Andrew Baird & Nolan Sawyer .... *Figure 13.11*

Andrew Baird & Nolan Sawyer .... *Figure 13.12*

Andrew Baird & Nolan Sawyer .... *Figure 13.13*

Andrew Baird & Nolan Sawyer .... *Figure 13.14*

Andrew Baird & Nolan Sawyer .... *Figure 13.15*

Andrew Baird & Nolan Sawyer .... *Figure 13.16*

Andrew Baird & Nolan Sawyer .... *Figure 13.17*

Andrew Baird & Nolan Sawyer .... *Figure 13.18*

Andrew Baird & Nolan Sawyer .... *Figure 13.19*

Andrew Baird & Nolan Sawyer .... *Figure 13.20*

Andrew Baird & Nolan Sawyer .... *Figure 13.21*

Andrew Baird & Nolan Sawyer .... *Figure 13.22*

Andrew Baird & Nolan Sawyer .... *Figure 13.23*

Andrew Baird & Nolan Sawyer .... *Figure 13.24*

Andrew Baird & Nolan Sawyer .... *Figure 13.25*

Andrew Baird & Nolan Sawyer .... *Figure 13.26*

Andrew Baird & Nolan Sawyer .... *Figure 13.27*

Andrew Baird & Nolan Sawyer .... *Figure 13.28*

Andrew Baird & Nolan Sawyer .... *Figure 13.29*

Andrew Baird & Nolan Sawyer .... *Figure 13.30*

Andrew Baird & Nolan Sawyer .... *Figure 13.31*

Photography Citations

Andrew Baird & Nolan Sawyer •••••••••••• *Figure 13.32*

Andrew Baird & Nolan Sawyer •••••••••••• *Figure 13.33*

Andrew Baird & Nolan Sawyer •••••••••••• *Figure 13.34*

Andrew Baird & Nolan Sawyer •••••••••••• *Figure 13.35*

Andrew Baird & Nolan Sawyer •••••••••••• *Figure 13.36*

Andrew Baird & Nolan Sawyer •••••••••••• *Figure 13.37*

Andrew Baird & Nolan Sawyer •••••••••••• *Figure 13.38*

Andrew Baird & Nolan Sawyer •••••••••••• *Figure 13.39*

Andrew Baird & Nolan Sawyer •••••••••••• *Figure 13.40*

Andrew Baird & Nolan Sawyer •••••••••••• *Figure 13.41*

Andrew Baird & Nolan Sawyer •••••••••••• *Figure 13.42*

Andrew Baird & Nolan Sawyer •••••••••••• *Figure 13.43*

Andrew Baird & Nolan Sawyer •••••••••••• *Figure 13.44*

Andrew Baird & Nolan Sawyer •••••••••••• *Figure 13.45*

Andrew Baird & Nolan Sawyer •••••••••••• *Figure 15.1*

Andrew Baird & Nolan Sawyer •••••••••••• *Figure 15.2*

Andrew Baird & Nolan Sawyer •••••••••••• *Figure 15.3*

Andrew Baird & Nolan Sawyer •••••••••••• *Figure 15.4*

Andrew Baird & Nolan Sawyer •••••••••••• *Figure 15.5*

Andrew Baird & Nolan Sawyer •••••••••••• *Figure 15.6*